COMPENDIUM OF CARTOGRAPHIC TECHNIQUES

COMPENDIUM OF CARTOGRAPHIC TECHNIQUES

Editor in Chief

JAMES P. CURRAN

Associate Editors

KJELD BURMESTER

ARIE J. KERS

E. SPIESS

Published on behalf of the
INTERNATIONAL CARTOGRAPHIC ASSOCIATION
by
ELSEVIER APPLIED SCIENCE PUBLISHERS
LONDON and NEW YORK

ELSEVIER APPLIED SCIENCE PUBLISHERS LTD
Crown House, Linton Road, Barking, Essex IG11 8JU, England

Sole Distributor in the USA and Canada
ELSEVIER SCIENCE PUBLISHING CO., INC.
52 Vanderbilt Avenue, New York, NY 10017, USA

WITH 2 TABLES AND 213 ILLUSTRATIONS

© 1988 ELSEVIER APPLIED SCIENCE PUBLISHERS LTD

British Library Cataloguing in Publication Data

Compendium of cartographic techniques.
1. Cartography
I. Curran, James P. II. International Cartographic
Association
526

ISBN 1-85166-229-4

Library of Congress Cataloging-in-Publication Data

Compendium of cartographic techniques.

Bibliography: p.
Includes index.
1. Cartography—Methodology. I. Curran, James P.
II. International Cartographic Association.
GA102.3.C65 1988   526'.028   88-11197

Printed in Great Britain by Page Bros (Norwich) Ltd.

Foreword

It is a distinct pleasure to write a few comments about the ICA Standing Commission on Map Production Technology chaired by Kjeld Burmester of Denmark on the occasion of the publication of this book.

This volume contains descriptions of 85 different techniques used in map production. Each is well illustrated by the 213 illustrations, which appear along with the descriptions. The task of the compilation of materials alone deserves much praise. However, this volume was compiled by an international team of fifteen experts from eleven countries. The monumental task of coordinating, translating, editing, and standardizing terminology was a tremendous undertaking.

Although the book is not intended to be a comprehensive text book, it certainly will be an indispensable reference volume in the training of cartographers in the universities and technical institutions of the world. It will be of interest to all map production organizations and should serve as an historic milestone in the recording of the technology used in late 20th century cartography.

To the authors, editors, and to the Standing Commission of Map Production Technology I extend the thanks of the ICA Executive Committee for the long hours of work that have resulted in this Compendium.

J. L. Morrison
President
International Cartographic Association
1987

Preface

The field of cartography is very extensive. As a result the cartographer, in making maps, is involved in many different processes, ranging from the original map design, compilation, and production through to the printing of the map. This Compendium, therefore, is intended to provide a general outline of the most pertinent techniques employed in map production and reproduction.

The growth of cartography over the last several decades has been immense. The most significant aspect of this growth has been in the field of computer-assisted cartography. Although complete treatment of this rapidly expanding subject lies beyond the scope of the Compendium, two chapters serve as an introduction.

When the compendium was conceived, it was the first attempt by Commission 2 to produce a complete, balanced presentation of cartographic techniques. The first step was to identify techniques which have relevance to cartographers. Under the direction of Professor Ernst Spiess of Switzerland, a list was prepared and Commission 2 members, authorities in their respective fields, then selected subjects for which they would provide both text and illustrations.

In 1984, when most of the material for the subjects had been collected, the Commission set up an Editorial Committee to coordinate the various subjects. This Committee was headed by James P. Curran (Canada), with Kjeld Burmester (Denmark), Arie J. Kers (Netherlands) and Professor Ernst Spiess (Switzerland) assisting.

More than three years of planning, writing and editing have elapsed. A major task was to ensure that subjects had international application, and to provide continuity between the presentation of subjects. There comes a time when research and writing must end in order that publication may begin. It is hoped that the quality of effort on the part of all participants is reflected in the Compendium.

Commission 2 considers that one of its more important tasks is to collect, analyze and disseminate knowledge of cartographic technology. We are therefore pleased to present the Compendium and hope that it will be of benefit to cartography internationally.

Kjeld Burmester
Chairman, Commission 2
Map Production Technology

James P. Curran
Editor

Contents

Acknowledgements

A publication of this magnitude could not have been published without the efforts of many present and former members of Commission 2.

The Commission would particularly like to thank the following contributing authors:

K. Burmester	Denmark
R. Cuenin	France
J. P. Curran	Canada
M. C. Doherty*	Canada
N. G. Grant	Canada
T. Kanakubo	Japan
W. Leibbrand	Federal Republic of Germany
A. Makowski	Poland
M. Miksovsky	Czechoslovakia
A. Papp-Vary	Hungary
B. Palinkas*	Hungary
C. Palm	Sweden
G. Piket	The Netherlands
E. Spiess	Switzerland
E. Sasvari*	Hungary

* Non Commission 2 members

The Commission also wishes to thank the following individuals and organizations who supported the project and prepared the illustrations.

Eidgenossische Technische Hochschule, Zurich, Switzerland.

Douglas Jackson, Surveys and Mapping Branch, Energy, Mines and Resources, Ottawa, Canada.

P. J. Carroll and M. Kumar, Australian Institute of Cartographers, Western Australia Division.

CHAPTER ONE

Image Generation

In the preparation of a map, there are usually a number of creative processes involved by which a new graphic form is generated. This process of image generation can apply to individual map symbols that have to be shaped and structured, as well as to a whole map comprising a variety of map symbol types. In the latter case there already exists a basic geometric plan provided by the absolute location of each item to be mapped. In the image generation process, however, this geometry may be modified or generalized to a greater or lesser degree. All these images are created manually or by digital methods.

Every analogue image generation process basically entails the establishing of a contrast of lightness or colour between the background surface and the image to be created. This background surface or base may be of any colour or lightness, but it is essential that its surface be homogeneous, even and consistent and suited to receive an image. An image contrast is established either by the addition or removal of pigments on, or from, the coating of the base material, or by the chemical conversion of existing pigments into other dominant wavelengths. The different processes are distinguished by the nature of the pigments and their treatment, and the drawing tools or instruments used.

Some of the most commonly employed cartographic base materials for image generation are listed in Table 1.1, and information on their specific

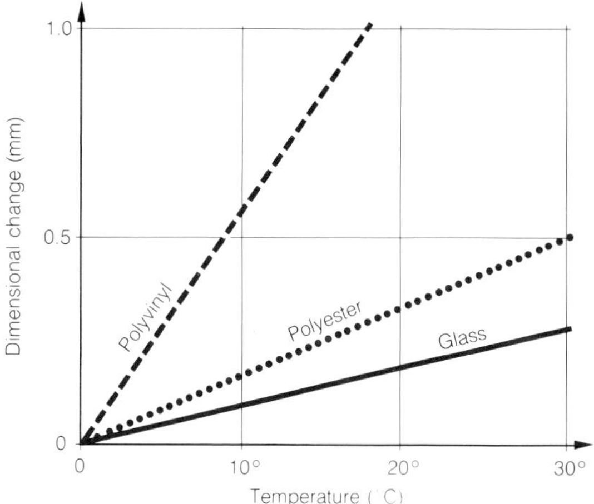

Fig. 1.1 Dimensional changes in various materials due to changes in temperature

properties and dimensional stability are shown in Figs 1.1 and 1.2.

1.1 Point Plotting

The plotting of points is the marking of exact geographical position on a map document, using a fine pricking instrument or sharp pencil in combination with other measuring instruments.

Point plotting is a common operation in map compilation. It is used for transferring single points from a source map or from a list of coordinates

Table 1.1 Characteristics of some of the most commonly used cartographic base materials

Base Material	Dimensional Stability	Transparency	Flexibility	Danger of breakage or tearing	Inflammability	Foldability	Ease of ink erasing	Other properties
Glass plates with laquered surface	excellent	high	none	easily broken	none	none	good with scraper	bulky
Metal plates laminated with drawing paper	very good	opaque	poor	none	none	none	poor with rubber	
Drawing paper	poor	opaque	very good	tears easily	fair	fair	poor with rubber	
Tracing Paper	poor	translucent	very good	tears easily	fair	fair	poor with rubber	Crumples when wet
Polyvinyl plastics*	good	highly translucent or opaque	good	splinters	high	low	good with scraper	Static electricity, melts at 55°C
Polyester plastics*	good	highly translucent or opaque	very good	none	high	high	good with scraper or wet rubber	unreceptive to dyes, melts at 150°C

* The surfaces of the polyvinyl and polyester plastics have been prepared during their manufacture to enable high quality ink lines to be drawn. For example, polyester has a grained or laquered surface.

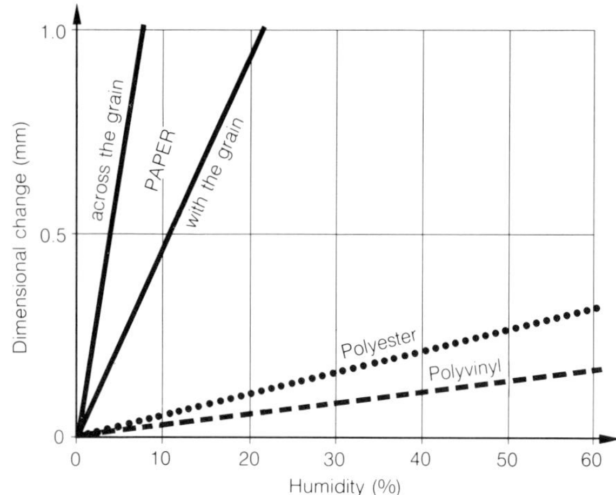

Fig. 1.2 Dimensional changes in various materials due to changes in humidity

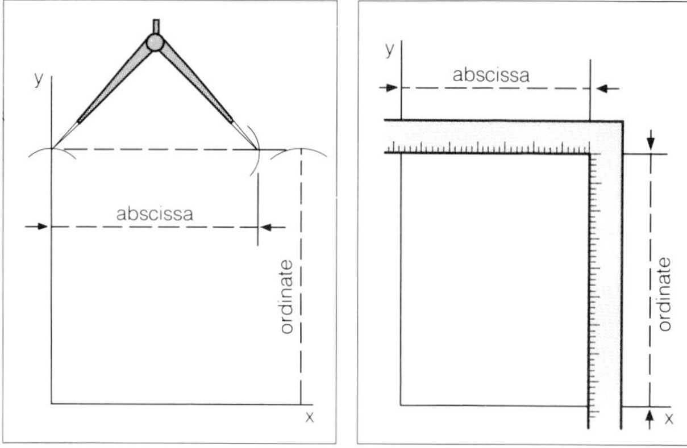

Fig. 1.4 Point plotting with (a) compasses or proportional dividers, (b) grid rulers or rectangular protractors

onto the map manuscript. The process may be used in the construction of a map projection or a grid in order to plot the graticule and grid intersections as well as reference points used to compile other map features.

The location of a point is determined with reference to two rectangular axes or two or more points already marked.

The precision to be expected under ideal manual plotting conditions is of the order of 0.1 mm. In addition to a pricker or pencil, other instruments that will be needed include:

—graduated rulers (Fig. 1.3).
—compasses, proportional dividers, grid rulers or rectangular protractors (Fig. 1.4 a and b).
—grid templates (Fig. 1.5).
—rectangular and polar coordinatographs (mainly used for highly accurate plotting) (Fig. 1.6).

The grid template, or master grid plate (usually metal) (Fig. 1.5), is used essentially for the pricking of grid intersections for grid construction. The

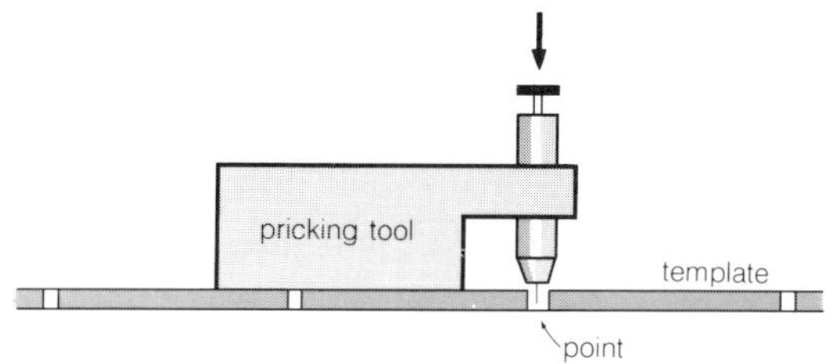

Fig. 1.5 Point plotting with a grid template and pricking device

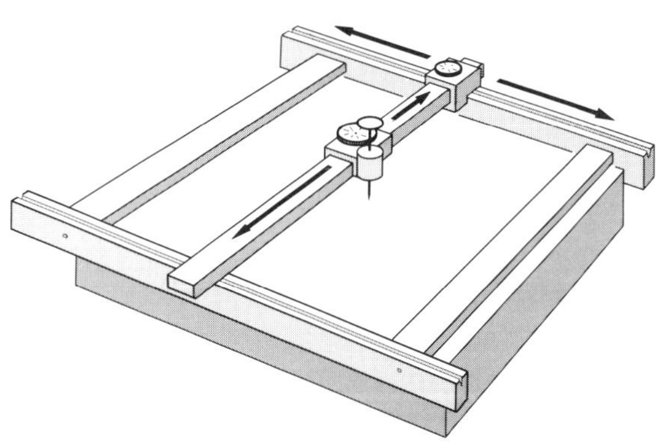

Fig. 1.6 Rectangular coordinatograph for manually plotting points

plotting of points is carried out through standard size holes arranged in a regular grid pattern, using a special pricking device.

For plotting a single point on a manuscript, a sheet of tracing paper, or plastic, overlaying the source material can be used and onto which the location of a point X is plotted along with three surrounding reference points A, B and C. The lines XA, XB and XC are drawn on the overlay which is then placed over the compilation manuscript so that these lines pass through A, B and C respectively. Point X is then pricked through the tracing paper onto the manuscript (Fig. 1.7).

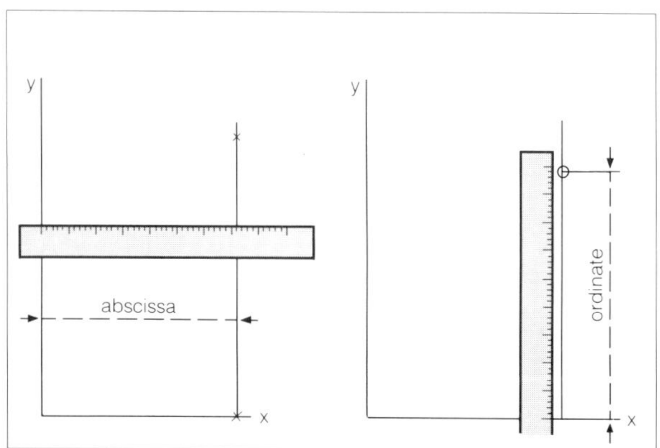

Fig. 1.3 Point plotting with graduated rulers

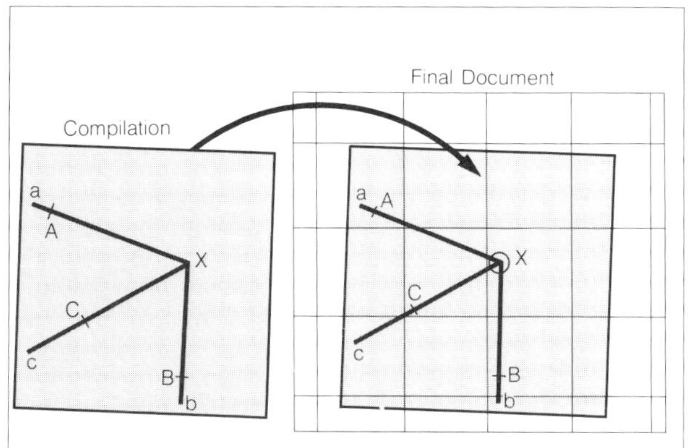

Fig. 1.7 Plotting a point in relation to three other points

For plotting a series of points, polar and rectangular coordinatographs are used. The rectangular version of these instruments (Fig. 1.6) has a gantry for the y-scale that moves at right angles to the x-scale and carries the slide with the pricking tool. The tool is positioned manually by setting the respective coordinates on the two scales before pricking the point.

1.2 Drawing Techniques

When lines are drawn with a pencil or pen, they are generally drawn from left to right along the edge of a ruler. When using ink, care should be taken to avoid the ink spreading between the base material and ruler. For drawing vertical lines, the base material is, where possible, rotated so that the horizontal drawing position is maintained. The drawing tool should be inclined about 10° so that the eye can follow its tip exactly and be guided steadily along the ruler edge with even pressure on the drawing surface.

Drawing with contour pens and road pens is normally executed with the hand free of the base material.

1.2.1 Line Drawing

The drawing of a line consists of depositing a pigment or another opaque substance with a pencil or other instrument onto an opaque, translucent or transparent surface. The line must have a uniform density that contrasts with that of the base material.

Line drawing is extensively used in map compilation as well as in the fair drawing of map reproduction masters. The drawing technique is applied mainly for linear map elements and point

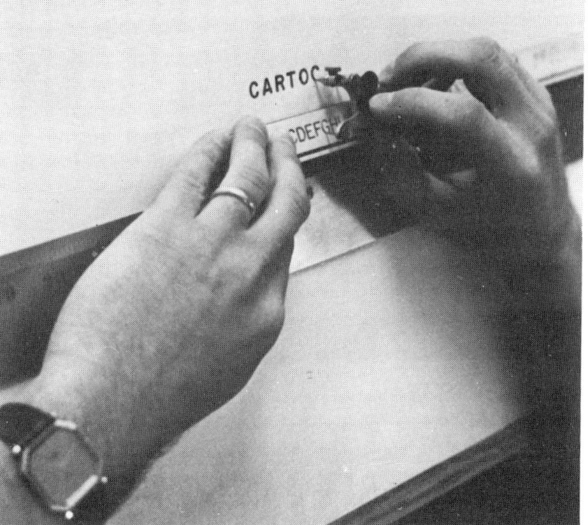

Fig. 1.8 Lettering instrument

symbols which are unique or occur rather seldom. It has certain advantages especially when the symbolization is of an individual character, e.g. pecked and dotted lines; or for the drawing of rock, sand-dune features or other irregular patterns.

Place and feature names can also be hand lettered directly onto a manuscript using fine pen nibs or specially prepared tapered brushes. However, this is a highly skilled and slow procedure; consequently, it is rarely used today. Lettering can also be applied with lettering stencils and ink reservoir pens with points of different widths that produce consistent weights of lines (Figs 1.8 and 1.9).

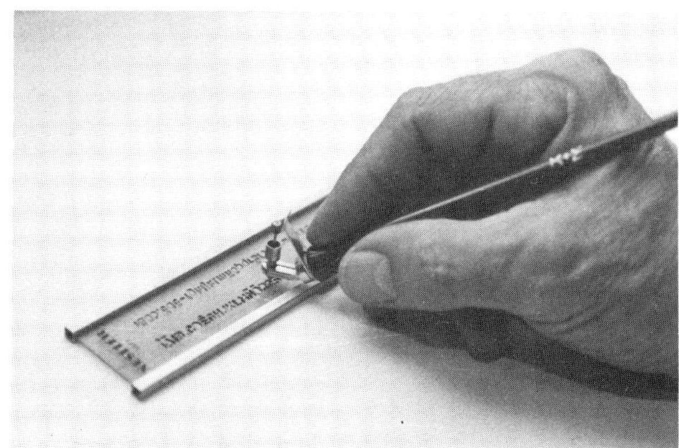

Fig. 1.9 Lettering stencil

The tools used for cartographic drawing include:

—pencils
—penholders with a large variety of pen nibs
—reservoir pens with a range of different line widths (Fig. 1.10).

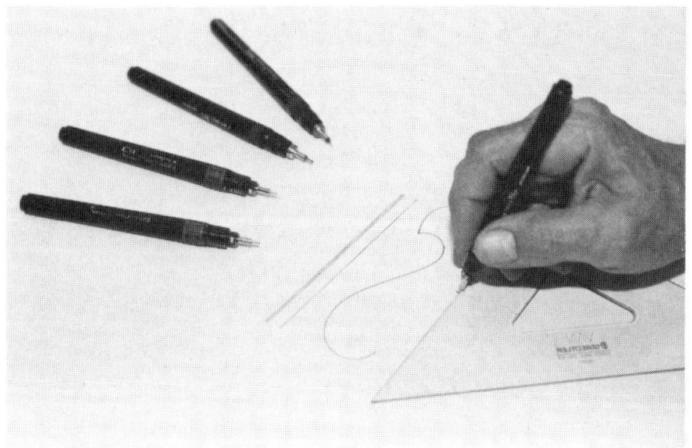

Fig. 1.10 Reservoir pens

—various ruling pens for single and double lines
  and swivel pens for contours (Fig. 1.11)
—drop and bow compasses for small circles
—brushes
—hard steel scrapers for corrections and line
  interruption.

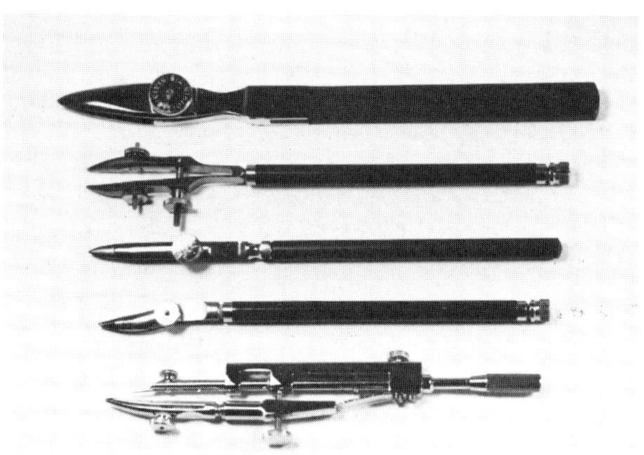

**Fig. 1.11 Ruling pens for single, double and curved
lines**

The use of these tools may require guiding instruments such as:

—rulers or straightedges,
—set squares, railway curves, French curves and
  adjustable curve rulers (Fig. 1.12).

When drawing with pens, the type of ink to
be used will depend on the characteristics of the
surface on which the ink is to be deposited, as a
variety of inks have been specially developed for
use on paper or polyester and polyvinyl plastics.
The fine pigment parts of these inks are distributed
in a solvent that evaporates rapidly when deposited
onto the base material. Thus, the ink dries and
the pigments adhere to the surface of polyester
plastics, or penetrate the drawing paper or polyvinyl plastics. Some inks are water soluble and can,

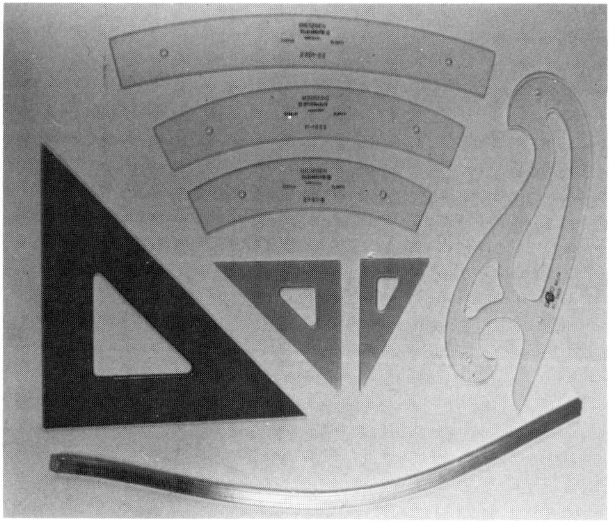

**Fig. 1.12 Set squares, railway curves and adjustable
curve ruler**

therefore, be easily removed from a plastic base
material with a wet rubber eraser.

1.2.2 Drawing of Solid Areas

The drawing of solids can be done to prepare masks
for areas that are to contain either solid colours,
screened colours, or patterns on the printed map.
The drawing or painting is done on a translucent
or clear sheet of plastic that contains either a
photomechanically prepared guideline image
copied from scribed line details, or the area outlines which have been traced in pencil by overlaying the plastic on the manuscript. The former
method, of course, is preferable for complex areas
in terms of accuracy. The drawing is done using a
photographic opaque or a special ink that is also
opaque.

The areas which are to contain a solid, tint or
pattern, are painted solid in positive form (Fig.
1.13). The completed drawing, if necessary, can

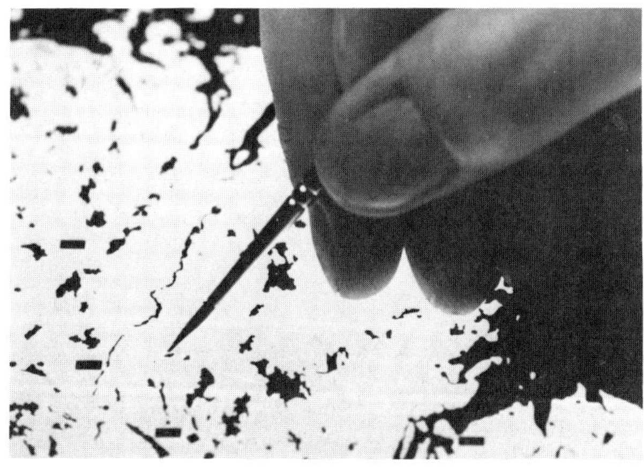

Fig. 1.13 Preparing a positive mask

Fig. 1.14 Contact negative mask

be photographically contacted to negative form (see 5.1) for use as a negative mask at a subsequent stage of reproduction (Fig. 1.14).

With the availability of materials for manually cutting and peeling masks (see 1.4) and photo-mechanical methods (see 5.10), the drawing or painting of solid areas by hand is very seldom done today.

1.2.3 Tinting

The plates containing tints are prepared to cover certain areas with a uniform gradation. They give the appearance of a continuous pale colour, differing from solids, which consist of saturated pure colours. Manually, tints are obtained by decreasing the colour saturation with the addition of white or some other solvent to dilute the colour.

Tints in monochrome images are a means of symbolizing an ordinal scale of areal features. These tints can be changed into another colour hue at a later stage of reproduction. (See 6.3.)

To paint tints homogeneously over each area is a highly skilled operation. Therefore, graded tints are now produced by other techniques, especially by screening (see 7.2). A visual effect of tinting can be obtained by drawing a fine regular pattern of cross-hatched horizontal, vertical and oblique parallel lines with a pen and ruler (Fig. 1.15). In fact, this is nothing but a line drawing that when seen from a certain distance gives the illusion of a tint due to the mixture of solid lines and the white spaces between them. The same effect can also be achieved using various size dot patterns.

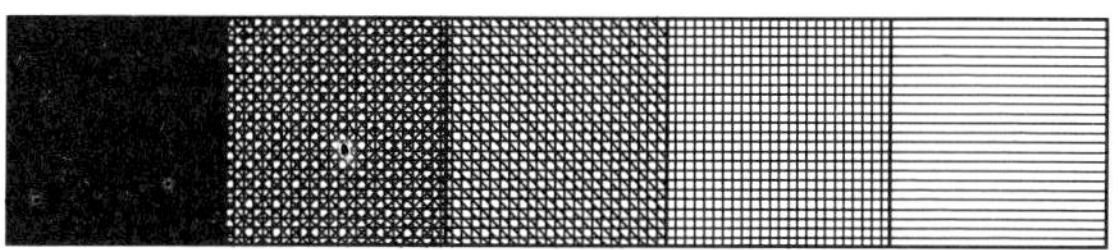

Fig. 1.15 Cross-hatched graded tints

1.2.4 Continuous-Tone Drawing

A continuous-tone drawing is one that contains the whole range of tones between black and white or dark and light. It can be produced in monochrome or full colour by applying more or less pigment to the dark and light areas respectively.

The main applications of this technique in cartography are: hill shading for the representation of relief and vignetting to enhance boundaries or outlines of areas. Both techniques are described in the following sections (1.2.4.1) and (1.2.4.2).

1.2.4.1   Shading

The technique of shading aims at creating a continuous-tone image of the Earth's surface or of a three-dimensional symbol on a map.

Techniques include: shading with black and white pencils on a sheet of white translucent plastic that has been uniformly coated with a grey tone, and on which the dark tones are drawn with a black pencil and the light tones with a white pencil. The light tones, or highlighting, can also be achieved by removing the grey coating by scraping with a sharp blade or with an electric eraser to expose the white plastic base.

The most common methods of producing monochrome and coloured shaded relief drawings are by shading with pencils and grey paper stubs (Fig. 1.16) used to smooth the drawing; by applying ink or finely mixed paint with an airbrush (Fig. 1.17) or with watercolour brushes (Fig. 1.18). Some of the materials on which drawings can be rendered include matt surface bromide paper, fine grained drawing paper mounted on, or laminated with a metal plate, and translucent matt surface white or clear plastics. If the latter medium is used, the light tones or highlights are normally applied to the face of the drawing with a white pencil or with a superfine gouache paint applied with an airbrush.

Fig. 1.16 Smoothing pencil drawing with a grey paper stub

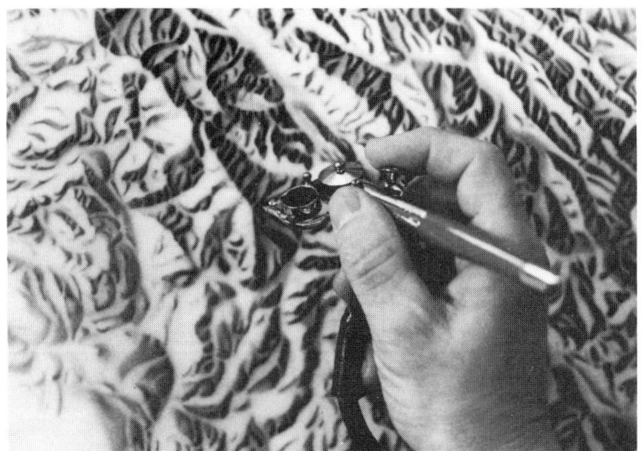

Fig. 1.17 Shading with an airbrush

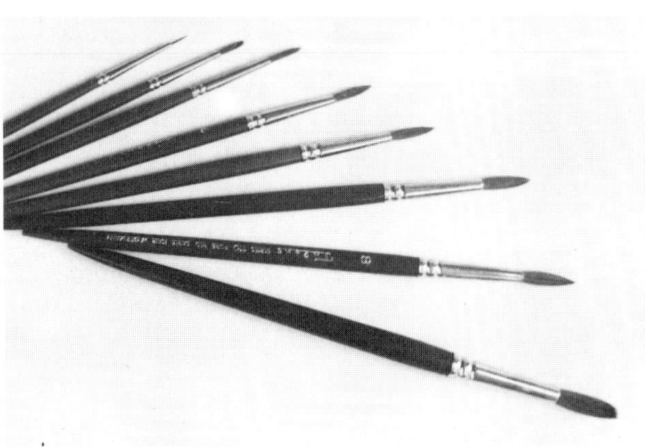

Fig. 1.18 Water colour brushes

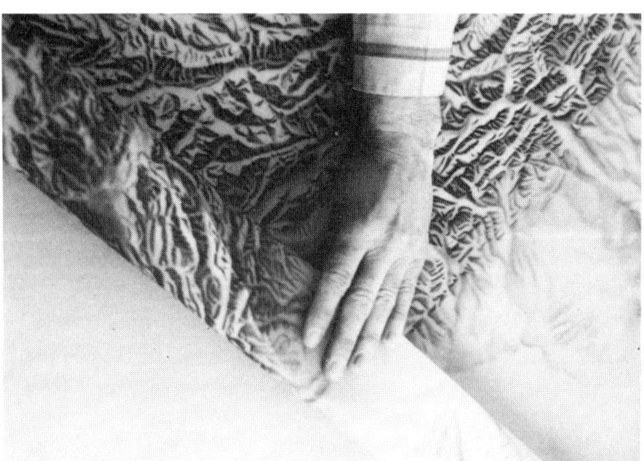

Fig. 1.19 Shaded relief drawing superimposed over sheet of grey paper

Fig. 1.20 Vignetted coastline

It is most commonly used to accentuate and enhance features such as coastlines (Fig. 1.20), lakes, wooded areas, glaciers and air traffic control areas, as it provides a clear outline for these features without detracting from the legibility of other map detail.

Vignetted bands can be produced by manually shading the area along the bounding line with pencils, stubs, brushes or air brush (Fig. 1.21). However, this method can be rather time consuming. Fortunately, vignetting can also be economically and effectively produced by photo-mechanical means (see 7.5).

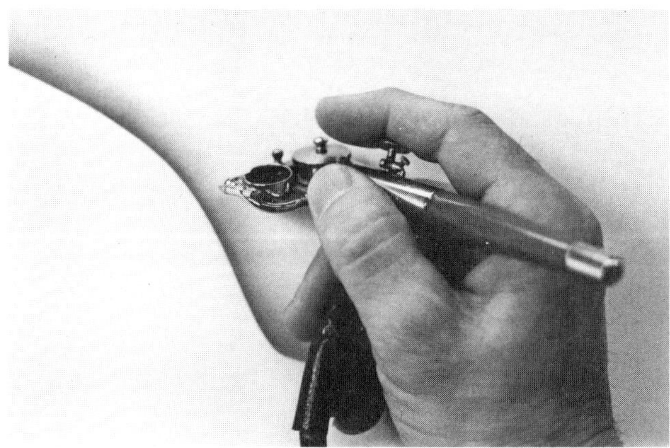

Fig. 1.21 Vignetting with an airbrush

The completed drawing is then superimposed on a sheet of grey paper (Fig. 1.19) and photographed in half-tone (see 7.1.1).

1.2.4.2 *Vignetting*
Vignetting, in the cartographic sense, refers to a band of colour used to emphasize a line bounding an area in which the tone gradually decreases away from the line.

1.3 Scribing

Scribing is the standard method of cartographic drafting today because it produces consistent high quality line work and is both easier and faster than traditional pen and ink drawing.

Scribing is done on transparent stable base polyester plastics or glass plates that have been coated with an actinically opaque engraving emulsion.

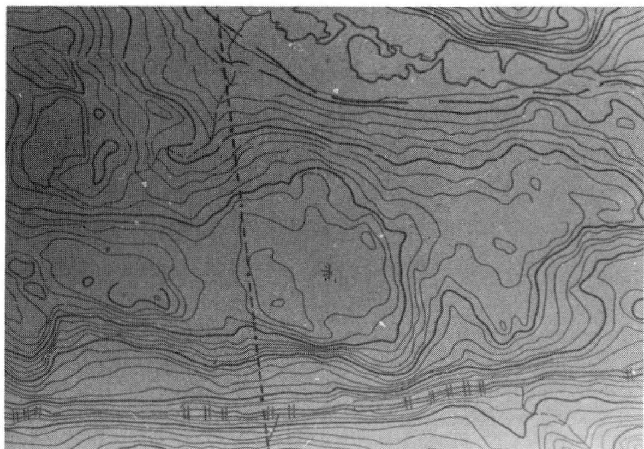

Fig. 1.22 Guide line image

The scribecoat plastics are available commercially and come in different sizes with different colour coatings, all of which have their own individual properties that will suit particular cases. The use of coated glass for scribing is limited, mainly because it is breakable, difficult to store and is not readily available commercially in different sizes. However, it has a very high dimensional stability.

The two principal types of scribecoat materials in use at present are commonly known as 'orange or rust scribe' which is semi-transparent and 'rust or red scribe' which is transparent. Orange or rust scribe is normally used when photomechanically prepared guide line images of the manuscript are required in order to scribe the map details (Fig. 1.22), whereas, the red scribe, because of its transparent property, is generally used to scribe overlay information. The scribing methods used by different cartographic organizations, of course will vary. For instance, the line details may be scribed in reverse or wrong-reading to produce a negative image (Fig. 1.23) which can be used directly to make a surface offset printing plate (see 5.10); or, the scribing may be done to produce a right-reading negative which can be contacted to make a reverse

or wrong-reading positive from which a deep etch printing plate is made (see 5.10). If errors are made in scribing, the unwanted line can be opaqued by applying a retouching solution which when dry can be rescribed.

Two other scribing materials which have found limited favour, but which should not be ignored, are 'white scribe' on which the scribing can be done with or without the use of a light table, and polyvinyl material with a dye-resistant coating used for positive scribing. When scribing is done on the white scribe material, the place and feature names can be applied to the white scribing surface. The material can then be mounted on a sheet of black paper so that the scribed image will appear black against the white scribing surface. The combined scribed image and names can then be photographed in the normal manner. When scribing is done on dye-resistant coated polyvinyl materials, the finished engraving is converted into positive form by spreading an etching ink over the surface of the material. The ink penetrates the base material in the area where the scribing has removed the coating to form a stencil. When the lines are sufficiently opaque, the dye-resistant coating is chemically removed carrying the surplus ink with it leaving a positive image of the original scribed detail.

Scribecoat materials are available in many different colours, e.g., yellow, blue, green and grey. A double-layered white scribecoat is also available on which the white scribing surface is coated over an actinically opaque rust coloured surface, thus making it suitable for contact copying. During scribing, both layers are cut away.

The basic scribing instruments in common use are the rigid, swivel and ring engravers. The rigid engraver (Fig. 1.24) is used for all single lines of uniform width such as contours and shore lines. The scribing points, ranging in size from 0.08 mm

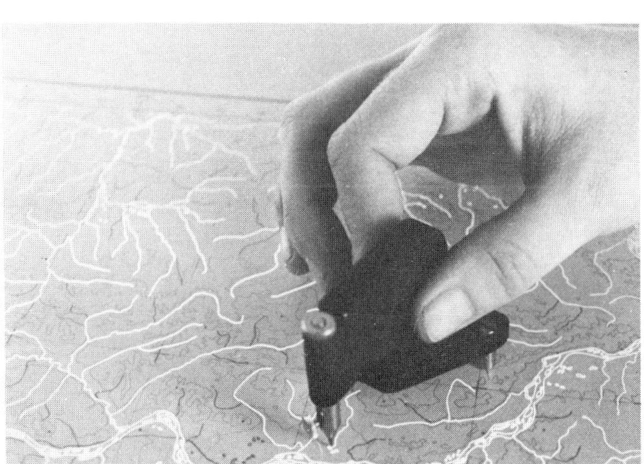

Fig. 1.23 Scribing line details

Fig. 1.24 Rigid engraver

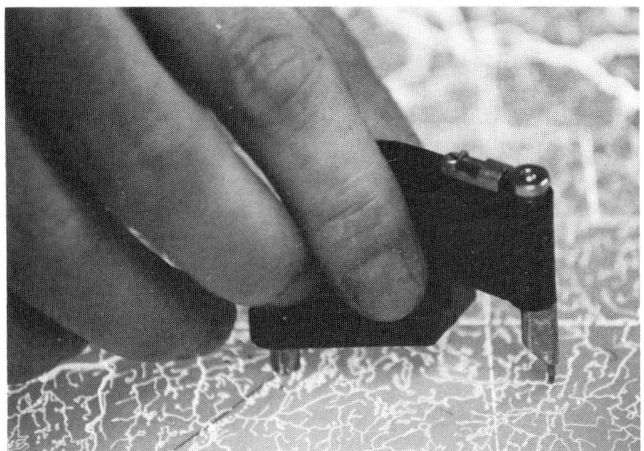

Fig. 1.25 Swivel engraver

to approximately 2.0 mm, are held in a socket on the cutting arm with a screw and are readily interchangeable when a different width line is needed. The swivel engraver (Fig. 1.25) is used to scribe double lines such as road casings, although for straight stretches the swivel can be locked and used with a straight edge.

Fig. 1.26 Subdivider

Fig. 1.27 Electric dot engraver

Other instruments designed for special purposes include sub-dividers (Fig. 1.26), electric dot engravers (Fig. 1.27), triple and quadruple line cutters, and an assortment of symbol templates (Fig. 1.28).

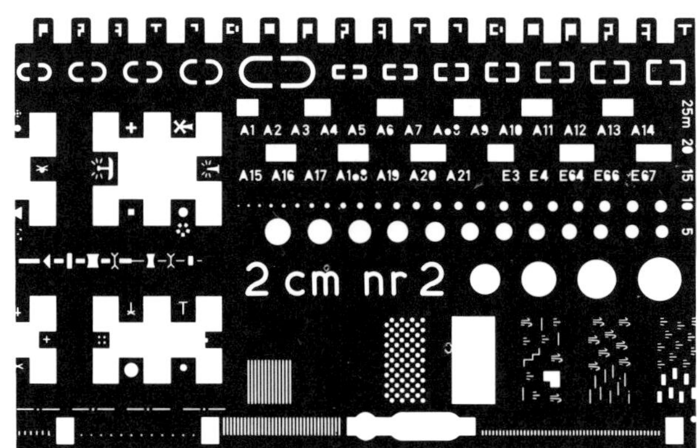

Fig. 1.28 Portion of a scribing template

1.4 Cut and Peel Technique

The cut and peel technique is one of several methods that can be used to make masks for areas which, on the printed map, will contain either a solid colour, screened tint or pattern. As the name implies, it is a method whereby a pliable film membrane can be cut and stripped away from its plastic base to produce either a negative or positive mask. A negative mask is produced when the image area is peeled out and, conversely, a positive mask is produced when the non-image areas are peeled out.

The material (cut and strip) is available commercially in all sizes compatible with other drafting, scribing and photographic materials. Its transparent polyester base has a pliable red stripping surface membrane which, although actinically opaque, is sufficiently transparent to see through; as a result it can be superimposed over any pencil, ink, scribed or photographic document for tracing or cutting.

Because of its pliable nature, the stripping surface cannot be cut with a rigid scriber. It must be cut with a knife (Fig 1.29) or with a ring or swivel engraver containing a needle or blade sharpened like a knife, after which it can be cleanly peeled away from its supporting base with the aid of tweezers or adhesive tape (Fig. 1.30).

As with all manual masking operations, great care must be taken to follow the outline keys exactly during the cutting process, as more often than not, the tint will have to fit a bounding line

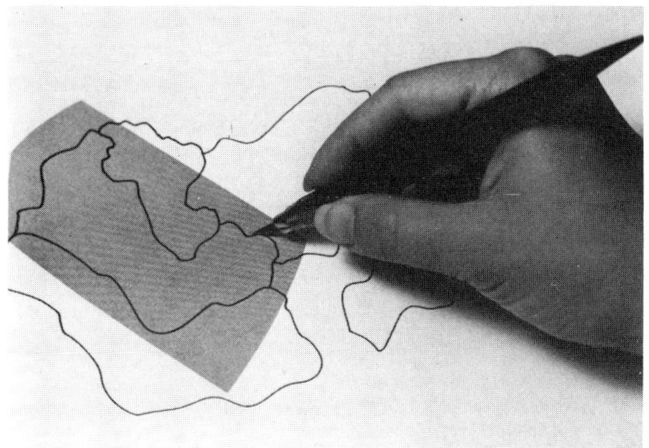

Fig. 1.29 Cutting surface membrane

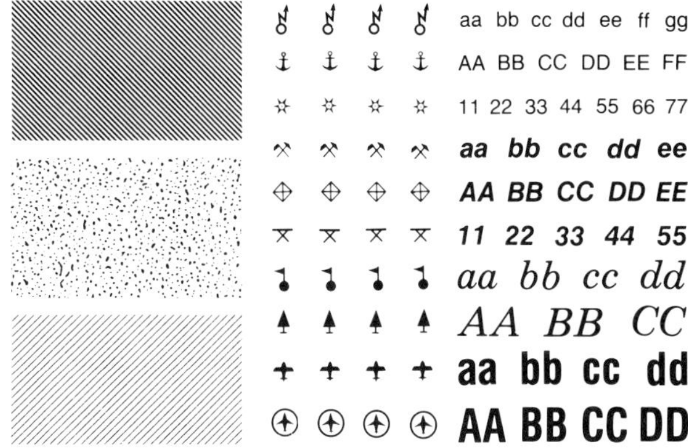

Fig. 1.31 Patterns, symbols and lettering for dry transfer

or abut another colour. Consequently, the technique is not normally used to produce masks for complex areas where precise registration is of the utmost importance, as in the case of shorelines, double line rivers, etc. (see 5.9). However, it is an effective method for producing masks for relatively simple shaped areas.

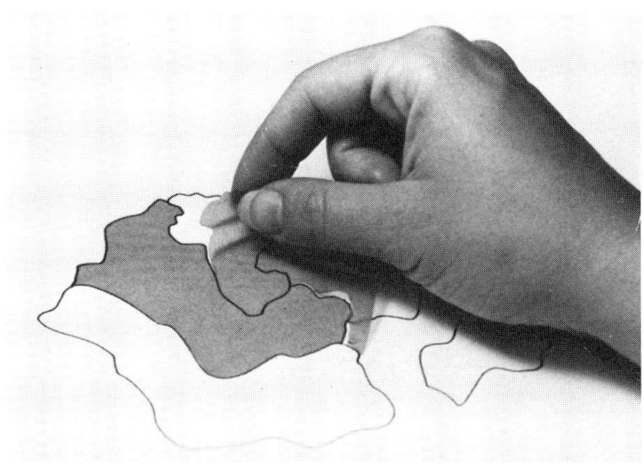

Fig. 1.30 Peeling away membrane

1.5 Dry Transfer Technique

The use of dry transfers makes it possible to apply lettering and symbols, in positive form, to various map components, and to combine line work with dot and other types of patterns or rulings without the aid of complex masking and photomechanical operations.

The pre-printed sheets are manufactured under a number of different trade names and are available in a wide range of patterns, symbols and lettering (Fig. 1.31). Because of the relatively small format of the sheets (25 cm × 36 cm approximately), the pattern screens are normally only used to cover small areas. However, with care, they can

be matched and joined to cover fairly large areas. The choice as to their use instead of photomechanically prepared tints or patterns, will depend entirely on the size of the final product, the required standard of quality, availability and cost.

Dry transfers are applied by first removing the protective backing. The letter or symbol is then transferred to the surface of the document by gently rubbing over the back of the transfer sheet with a blunt instrument or soft pencil (Fig. 1.32) and then peeling it away, leaving the letter or symbol positioned on the document (Fig. 1.33). The same sequence is repeated when individual letters are used to make up place and feature names.

Mistakes can easily be removed by lightly pressing a piece of adhesive tape onto the letter and lifting it away from the surface of the document. The completed work can be sprayed with a clear protective coating to prevent abrasions from handling or storage.

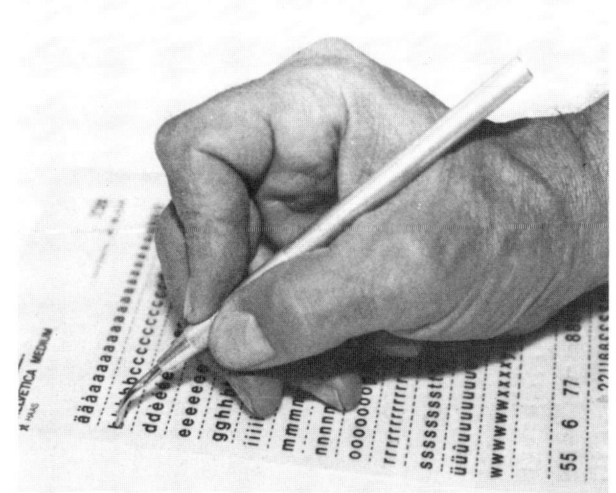

Fig. 1.32 Transferring letters to documents

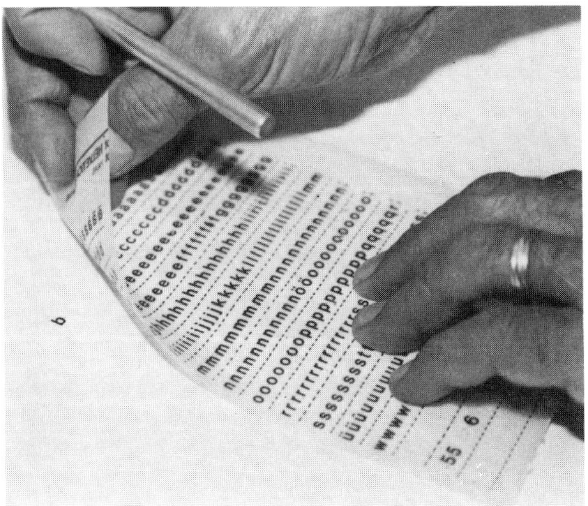

Fig. 1.33 Peeling away transfer sheet

Presensitized transfer films are also available commercially on which one can make their own dry transfers. The film is exposed to a strong ultraviolet light source in a vacuum printing frame in which it is placed in direct contact with a negative containing the symbols or patterns. Following exposure, the latent image on the film is developed with a special developing solution, after which, the image can be transferred onto the map document in the same way as the pre-printed transfers.

For repetitive and standardized mapping operations, it can be advantageous to prepare one's own rub-on symbols. The most used symbols are drawn once, repeated photographically, and mounted on a transparent sheet of plastic or film. This film can then be copied onto the commercially available dry transfer film as required.

1.6   Stick-on Technique

This technique allows for an image transfer of point, line and area symbols, including lettering,

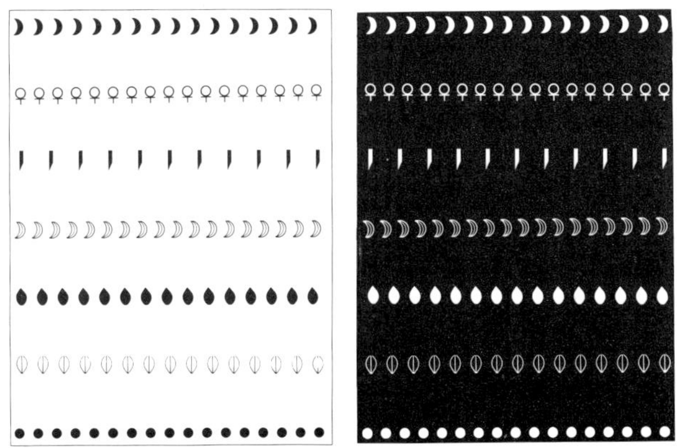

Fig. 1.34 Positive and negative stripping film symbols

to an opaque or transparent base material. The image to be transferred can be printed onto an intermediary base or photographically reproduced onto stripping film in either positive or negative form (Fig. 1.34). The latter is mainly used for additions or corrections to negative reproduction masters.

1.6.1   Stick-on Point and Area Symbols

Prior to application, the intermediary material carrying the image of the point or area symbols to be transferred is coated with an adhesive substance, usually wax (Fig. 1.35) and covered with a protective backing (Fig. 1.36) that is later removed before the image is applied to the base.

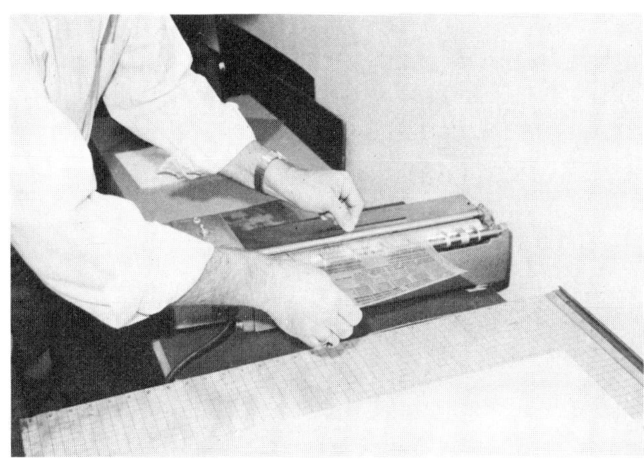

Fig. 1.35 Applying adhesive wax

The stripping film or sheet carrying the image can be coated either on the reverse or wrong-reading side in order that the symbols can be applied to the base in right-reading form or, on the right-reading side in order to be applied in wrong-reading form.

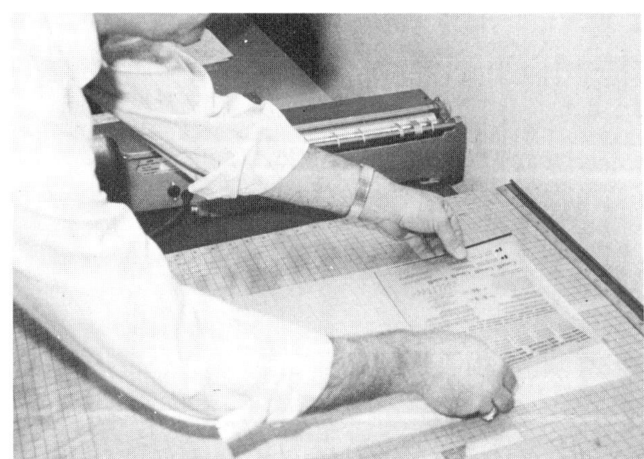

Fig. 1.36 Applying waxed material onto protective backing

To apply the pre-waxed name or point symbol, the image to be transferred is cut around its outlines leaving a small rim of extra material (this is to avoid complex retouching of the cut outlines which often appear on the subsequent negative reproduction masters). The protective backing is then removed to expose the wax adhesive and the name or symbol is placed in position and burnished with a smooth edge onto the base (Fig. 1.37).

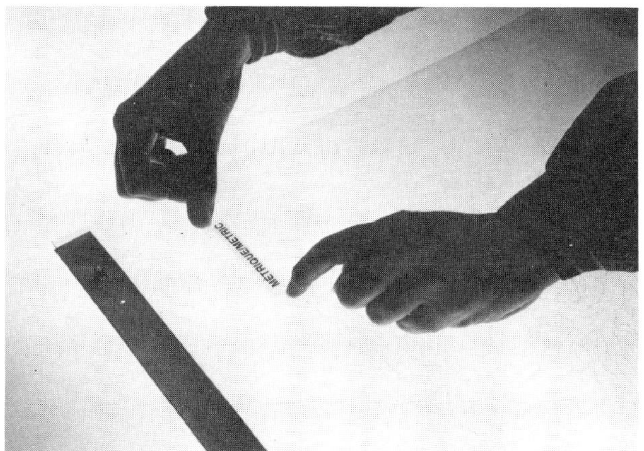

Fig. 1.37 Positioning name

In the case of area patterns or screens, the protective backing is removed. The pattern or screen is then applied to the base and cut with a sharp blade precisely to fit the outlines of the area (Fig. 1.38). The unwanted parts outside of this area are then peeled away (Fig. 1.39) and the remaining parts are carefully burnished onto the base.

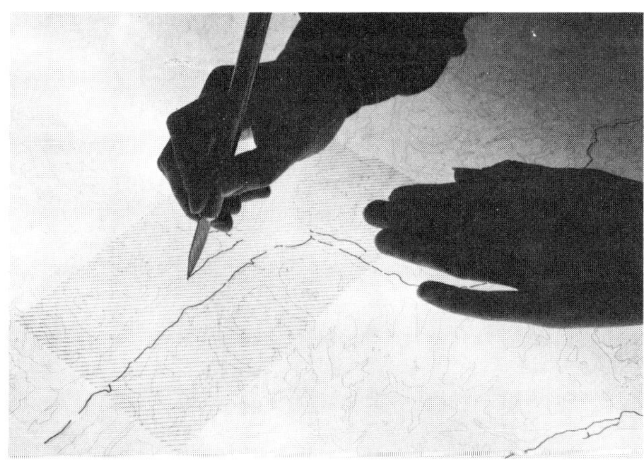

Fig. 1.38 Cutting area pattern to fit bounding lines

1.6.2 Stick-on Line Symbols

A wide range of line symbols are commercially available in the form of adhesive bands on rollers (Fig. 1.40). Narrow straight or curved lines are

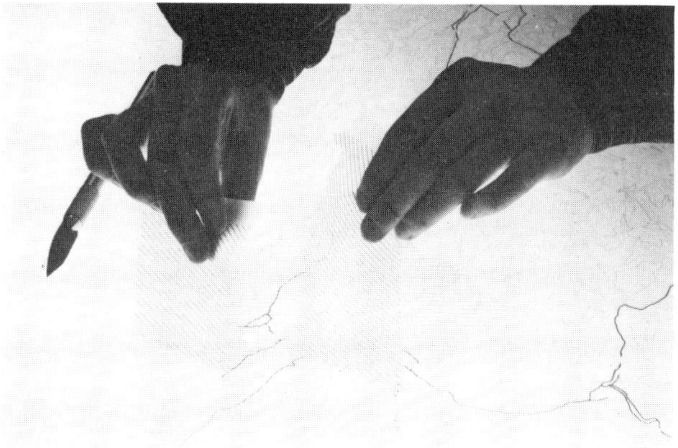

Fig. 1.39 Removing unwanted material

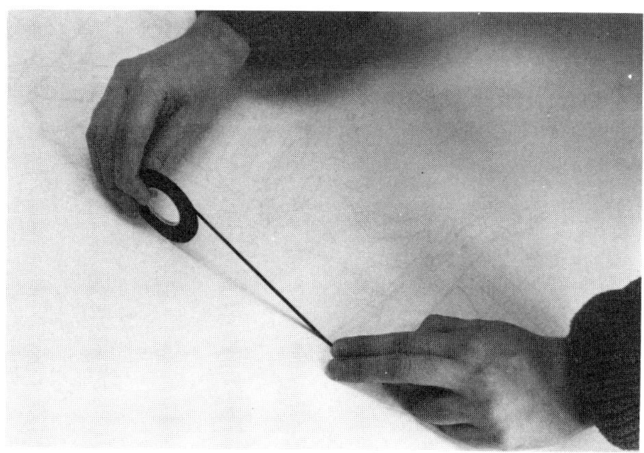

Fig. 1.40 Adhesive band applied directly from roll

usually applied with a special dispenser (Fig. 1.41). The lines can be placed manually or by precisely following a straight ruler or French curve. The rolls are available for black or coloured solid opaque lines including hachured or screened lines and repetitive symbols (Fig. 1.42) that can be applied to either opaque, translucent or transparent base drawing materials.

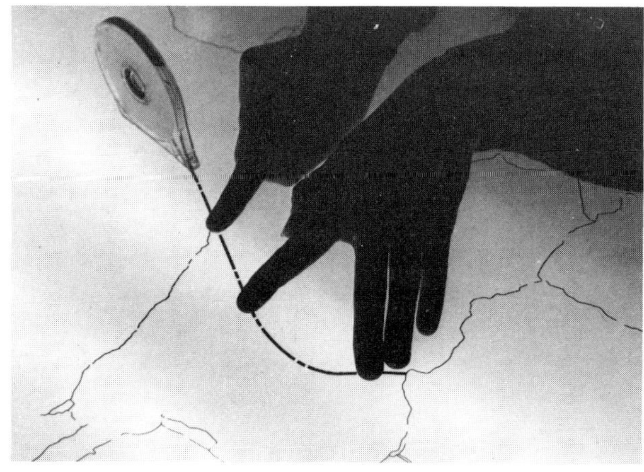

Fig. 1.41 Curved line applied with a dispenser

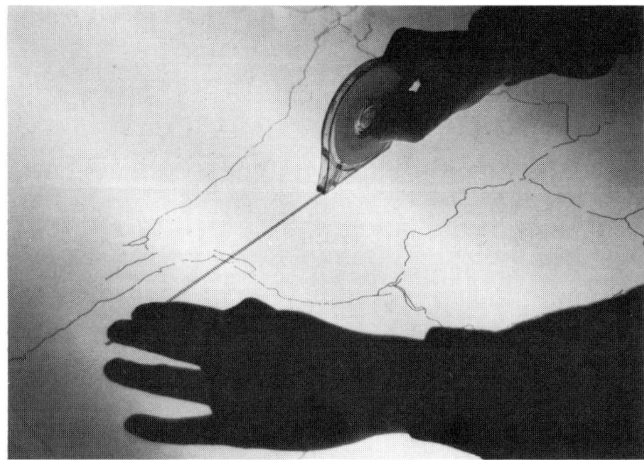

Fig. 1.42 Applying pecked or broken lines

The technique is useful for constructing drawings that are to be photographically reduced in scale, or for wall maps that require bold lines (Fig. 1.43).

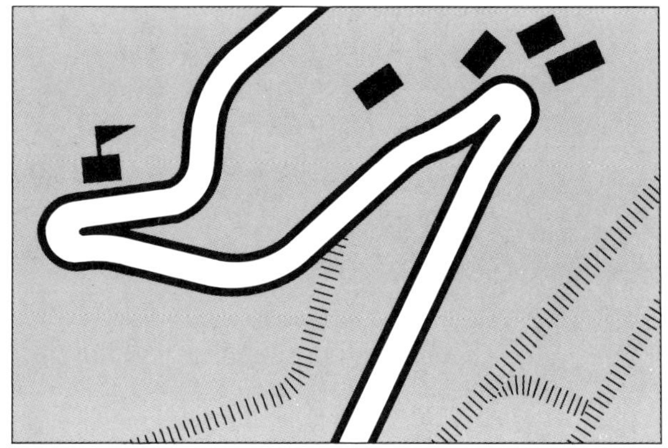

Fig. 1.43 Part of a wall map produced with adhesive bands

1.7 Lettering

There are several methods by which the lettering for place and feature names, as well as other textual information, can be added to a map document. The letters can be hand drawn directly onto the map manuscript; however, because this is a highly skilled and extremely slow process, it is very seldom used in cartography today. Lettering can also be drawn with the aid of mechanical lettering devices and lettering stencils using pencils or ink reservoir pens (see 1.2.1). This latter method, although sometimes used for lettering on large scale plans, is not applied in the production of topographic or thematic maps. Lettering can also be applied with commercially available dry transfer alphabets as described in section 1.5. Again, like lettering with the aid of mechanical lettering devices and stencils, this method is not regularly used in topographic or thematic cartography.

Before the advent of phototypesetting machines, type was manually hand set by a compositor from fonts of metal type characters stored in wooden trays that were divided into compartments for specific letters and numbers. The assembled type was inked and transferred to a thin white paper or clear film from which the names could be cut out and applied directly onto the manuscript with an adhesive substance.

Several mechanical typesetting machines were developed to remedy the slowness of manually setting type. The two principal types are the Linotype and Monotype machines, both of which, have been extensively used commercially to produce large quantities of text. However, they have had limited use in cartography, where, by comparison, relatively fewer names and texts are required.

The chief method of preparing place, feature names and text, in most cartographic organizations today, is through the use of phototypesetting equipment (see 1.7.1).

1.7.1 Phototypesetting

Phototypesetting is a process that makes use of a photographic technique and results in lettering on photographic film or paper. These names, or text, are then placed manually onto the reproduction masters (usually transparent plastic overlays) by the stick-on technique (see section 1.6).

The letters are typed on a keyboard or similar device, either 'off-line' or 'on-line' to the phototypesetting machine. In the former case, data are transferred by magnetic tape or disc to a computer for further processing. In the latter case the data is transferred directly to the phototypesetter where the input command for a certain letter

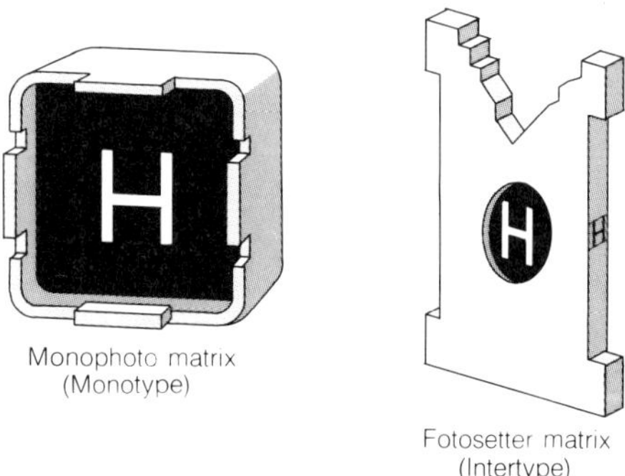

Fig. 1.44 Matrices of the first-generation phototypesetters

immediately initiates the exposure of the photographic material through the negative image of the required character. These negatives are contained on a rotating disc or matrix font that is stopped for each exposure.

The first generation of phototypesetting machines to be constructed on electromechanical principles was derived from similar mechanical typesetting machines, the hot-metal melting part of the equipment having been replaced by an arrangement with transparent photographic matrices (Fig. 1.44).

The second generation phototypesetting machines (Fig. 1.45), which use rotating discs or drum matrices, have a moderately high output speed. A control system allows for interletter spacing between each exposure and the character size can be changed with a zoom or sizing lens. These machines are usually accessed with an encoded paper or magnetic tape prepared on a separate keyboard and operated electrically by a process computer. In the simpler versions of these machines, the rotating disk matrix is operated by hand.

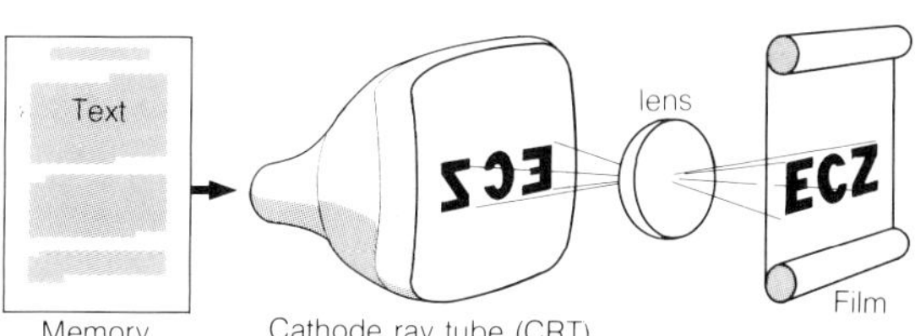

Fig. 1.46 Schematic layout of a third-generation phototypesetter

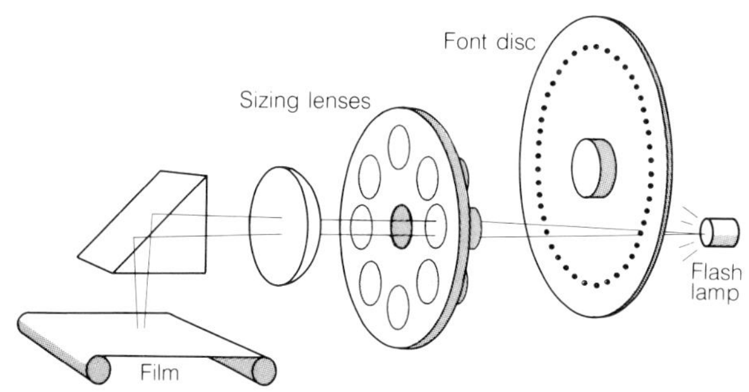

Fig. 1.45 Schematic layout of second-generation phototypesetter

Third generation phototypesetting machines (Fig. 1.46) are noted for their ultra high speed output. The type characters are stored in digital form either on a magnetic tape or disc and encoded as to type size and style. Each character is generated by a programmed arrangement of minute squares and flashed onto a high resolution cathode ray tube (CRT) and exposed through a lens onto photographic film or paper that is passed in front of the luminous CRT screen.

CHAPTER TWO

Digital Data Capture

digitizing and use the keyboard to change a feature code.

Manual digitizing at a graphics workstation (Fig. 2.6) is also used to edit and revise existing digital cartographic files when an update or revised map file is required.

Fig. 2.6 Graphics work station with large format digitizing table

2.3 Raster scan digitization

Raster scan digitization is an automatic process where the point, line and area features of a map can be converted to a digital representation using electro-optical, solid state, or laser sensors which traverse the map image in a regular series of scan lines.

The data produced by scanning represents a digital raster data structure which is different from the vector data structures produced by digital stereo compilation and manual digitizing systems.

The principles of scanners are the same whether they are rotating drum scanners (Fig. 2.7) or flatbed scanners (Fig. 2.8) with movable gantries. The entire surface of a map component, whether in negative, positive or full-colour form, is sensed in

Fig. 2.7 Rotating drum scanner

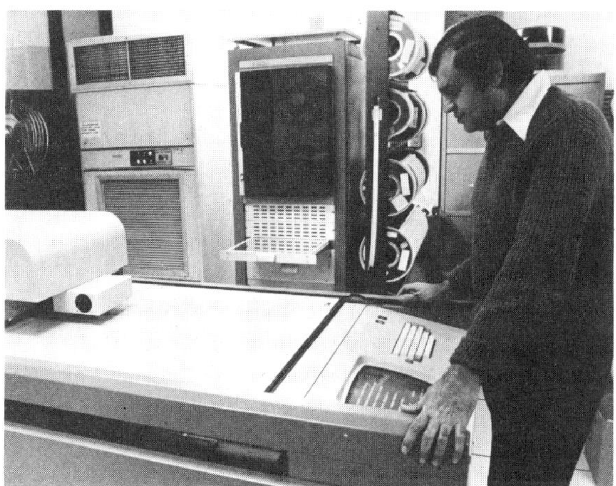

Fig. 2.8 Flatbed scanner

a series of parallel scan lines. These scan lines are further subdivided for recording purposes into smaller units of area, called picture elements or pixels. The sensor or the scanner head, records digitally the density or amount of light transmitted or reflected over each pixel, creating a matrix, or two dimensional array of discrete pixels over the surface of the scanned image (Fig 2.9).

At each pixel position, the sensor takes a reading, and can detect from 1 to 127 grey tones between black and white, for a monochrome document. The sensor can be calibrated to record the captured data as a binary raster image. For example, the presence of black (e.g. a line image) is recorded as '0'. If less than a set amount of the pixel area is black, the sensor records the pixel as a '1' value or white. The resolution of the scanner is determined by pixel size, and can be varied in a range between 25 to 200 μm.

Scanning produces large volumes of uncoded raster graphics which can be viewed unprocessed

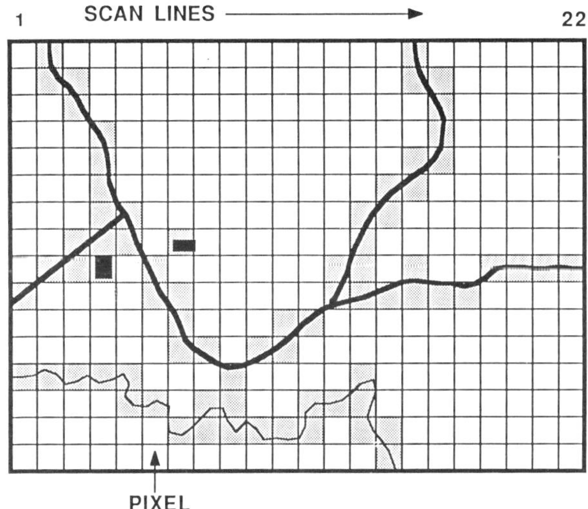

Fig. 2.9 Schematic of scan lines and individual pixels used to capture data

on a display terminal or output to a raster plotter to view the captured image.

It is often necessary to convert the raster data to vector form using specialized software in order to incorporate the data into a database and to manipulate it through the use of more conventional digital mapping systems. Once vectorized, the cartographic data can be manipulated on an interactive workstation or edit station, which will include the addition of cartographic feature codes to appropriately identify lines, points and area features.

2.4 Automatic line following (digitizing)

Automatic line following digitizers have a laser or photo-electric sensor head which can be locked onto a line (e.g. contour) on a negative. Once locked on to a line, the scanner follows automatically from its start to the end (e.g. the edge of the map sheet, a node or return to the starting point in the case of a loop), recording the X and Y coordinates of the line being digitized.

The map component to be digitized is reduced to negative small scale format. Once the scanner is locked onto a line, it can be feature coded, and the operator can observe the progress of the automatic line following on the display screen. If in the process of scanning, the sensor encounters a crossed line or some other ambiguity, such as a poor quality image, the operator can intervene, identify the correct line to follow, and the scanning will continue automatically.

Once a complete line has been digitized, it can be reviewed and eliminated from the operator's screen, leaving only those lines which have yet to be digitized. The cursor on the screen is then moved to the next line to be digitized and the process is repeated.

2.5 Digital stereo compilation

Photogrammetric stereo compilation is used to capture cartographic data directly from aerial photographs. It is unique among the methods of cartographic data capture in that the stereo model formed within the stereoplotter permits observation and capture of the third dimension (e.g. contours and spot heights).

Conventional stereoplotters produce a hard copy compilation of hydrology, cultural features and contours, etc. This is achieved by manually tracking the feature with a light projected plotting point or mark which is visible in the stereomodel

Fig. 2.10 Digital stereo compilation

created by a pair of overlapping aerial photographs on the stereoplotter. As the feature is traced, it is simultaneously drawn with a mechanical pantograph mechanism connected to the plotter onto the paper or plastic drafting film.

Stereoplotters can be retrofitted with digital encoders in several ways. The encoders can sense the rotary motion of spindles which the operator uses to move the measuring or tracking mark over the stereo-model, or the encoders can sense the motion of a tracing stand which the operator uses to position the measuring mark.

Another type of stereoplotter is the analytical stereoplotter. In this device the digital encoders are an integral part of the operation of the plotter, specifically designed to capture data digitally.

When a stereoplotter is fitted with digital encoders, the cartographic data can be captured and stored, through the use of a computer, on a mass storage device (e.g. magnetic tape or disk).

As with manual digitizing (see 2.2) the operator can assign feature codes to the cartographic data being captured by means of menu or via keyboard input (Fig. 2.10). The addition of feature codes permits sophisticated processing, manipulation and output of the digital cartographic data.

Some early digital stereoplotting systems performed 'blind' digitizing. Using this type of system, the computer captures and stores data, but does not provide the operator with an immediate visual display for the captured data. The display or hard copy plot of the data has to be generated after the stereoplotting session is complete.

Newer digital stereoplotting systems incorporate an interactive graphics screen. This provides the operator with an immediate visual display on a graphics monitor of the features being digitized, and the ability to interactively edit and manipulate data.

Image Registration

To ensure that individual colour components precisely fit each other on the printed map, it is necessary to provide colour registration marks. These marks provide a visual comparison of the registration and ensure the accurate fit of the various map elements throughout the successive production operations, from the initial drawing or scribing, through the camera and photomechanical operations to the printing of the map.

3.1 Registration Mark System

Many different types of registration marks have been developed to meet different requirements. They can be round or quadratic in shape and used in either positive or negative form (Fig. 3.1). The choice of which type to use is determined by the production requirements and facilities of each individual organization.

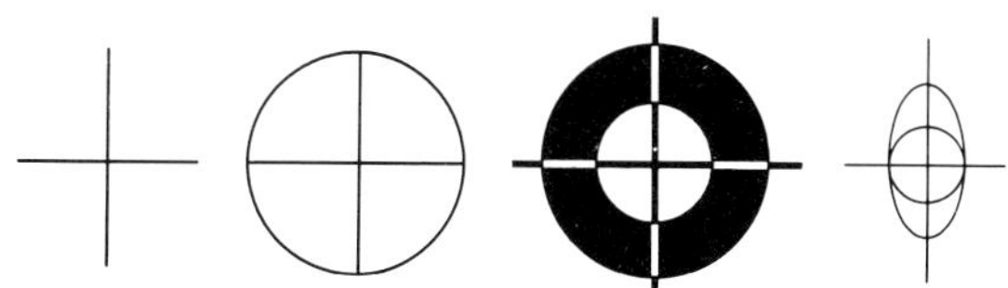

Fig. 3.1 Samples of various type registration marks

In the production sequence of a multicolour map, the registration marks must first be applied on the original source material, as all subsequent line elements, masks, colour plates for road fills, wooded areas, settlements etc. are registered to the same source material by placing the marks exactly on top of the original marks.

In most cases, four corner registration marks are either drawn in ink or scribed on to each component (Fig. 3.2). The marks can also be photographically reproduced on stripping film,

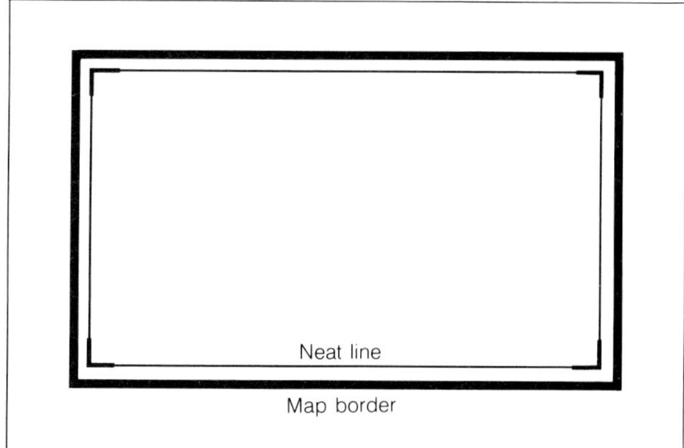

Fig. 3.2 Location of corner registration marks

usually in positive form, and stuck down over the four corners on the transparent plastic overlays.

Registration marks can also be located in the exact centre of all four sides of the components at right angles to each other. This is usually done if the components are to be photographed or duplicated by contacting (see 5.1). If eventually the printed map is to be trimmed to a certain size, they must be placed outside the trim area but within the printing format (Fig. 3.3).

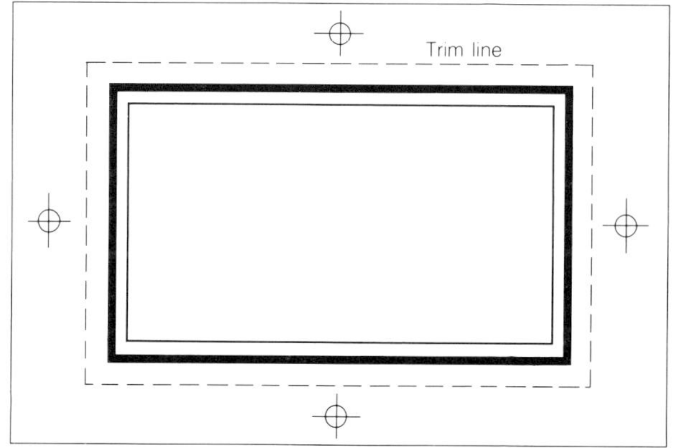

Fig. 3.3 Location of registration marks outside trim area of map

It is advantageous if the registration mark for the darkest colour is the smallest and that those for each successive lighter colour be slightly larger. This provides for better visual comparison of the registration between different colours.

In all cases, the marks are used to position the images onto the printing plates and to provide successive visual registration checks at the colour proofing, and printing stages (see 9.3 and 10.1).

3.2 Pre-punch Registration System

Because individual map components (e.g. scribed line work, colour masks, names etc.) are usually in negative form, it is extremely difficult to visually register them accurately on top of each other. To overcome this, a number of different pre-punch registration systems, with an arrangement for cutting holes or slots in film or plastics, have been designed to make it possible to pre-punch sets of repromat so that the material can be placed over studs, or onto a register bar containing fixed studs, in any combination with each other in perfect register (Fig. 3.4).

Registration systems can be classified as edge or central punch systems. In the edge registration

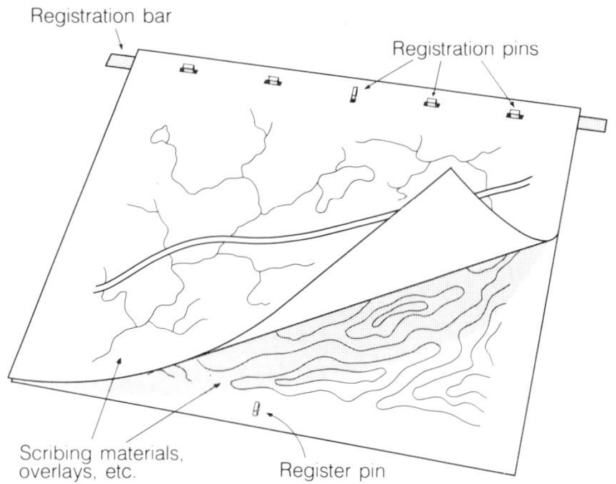

Fig. 3.4 Registered assembly

systems, the punch holes are collinear along one edge of the material (Fig. 3.5). Whereas in the central system the holes or slots are positioned along three or more edges of the material (Fig. 3.6). The choice as to which system to use will depend on the operational facilities and reproduction requirements of the organization.

The systems range from simple two or three hole punch devices for small format work to elaborate

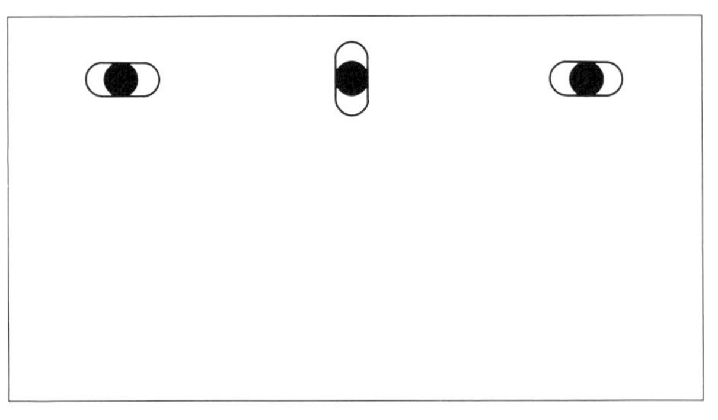

Fig. 3.5 Edge punch registration system

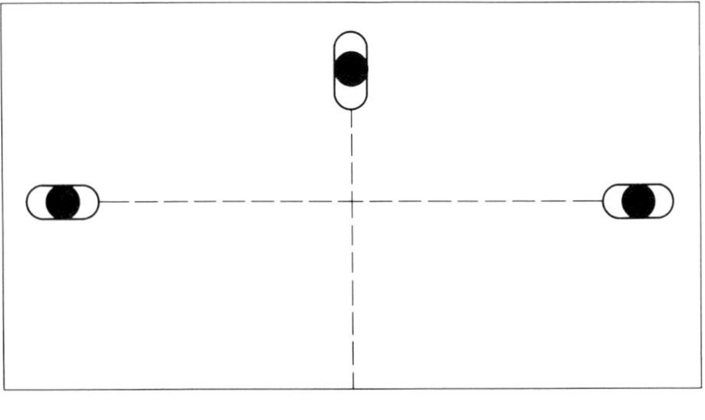

Fig. 3.6 Central punch registration system

five or seven hole precision punch units that can accommodate the largest size of commercially available materials. In either case, the system requires that the holes be punched at fixed distances sufficiently far enough apart in order to control any swing in the material. When large sheets of material are used, it is desirable that the punch system has the capability to punch a slot in the centre of the edge of the material directly opposite the 'gripper edge' (Fig 3.7). This provides additional control.

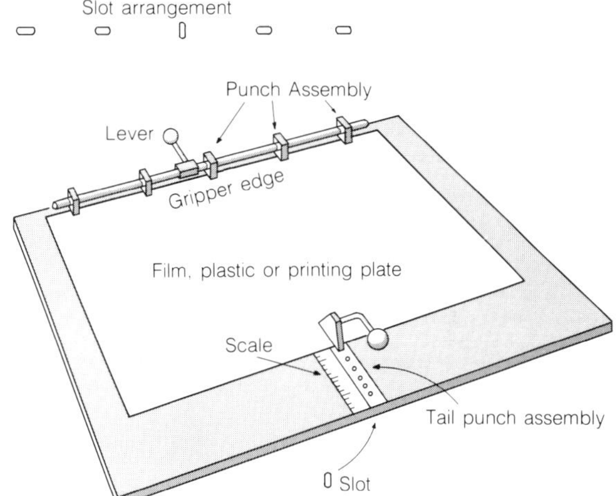

Fig. 3.7 Typical pre-punch registration equipment

The sequence of punching is for the first document in the production chain (usually the base manuscript) to be punched, after which all other job-related materials are prepunched prior to any work being done. This ensures the same registration for the various components throughout the production stage to the final printed map.

Some of the more elaborate punch units can be correlated with permanently installed pins on the plate cylinder of a printing press. This ensures that each printing plate is set on the cylinder in the same position. As a result, the pressman has very little adjustments to make, thus cutting down on the 'make ready' time.

3.3   Stick-on Pin Registration System

Plastic tabs with precision-punched holes, along with corresponding pins, provide an effective and economic means for registering complementary drafting and reproduction materials to allow them to be brought together in correct registration to each other at any stage during the production process of the map. The use of this system does not require access to a mechanical punch registration

Fig. 3.8 Tabs with pins inserted attached to longest side of material

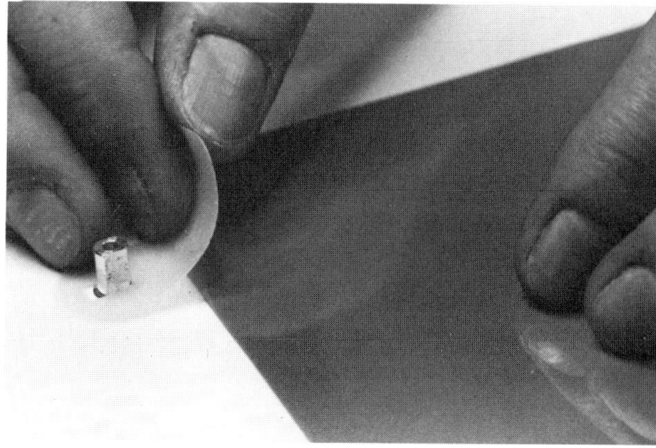

Fig. 3.9 Other related job components registered over first component

device. Both the tabs and pins are available commercially. The tabs have an adhesive backing and come in various sizes with different shaped pre-punched holes. The pins can be used individually or attached in series to a registration bar at set distances apart.

The sequence of registering the individual job components is to first attach two of the punched tabs to the longest side of the material at a suitable distance apart to provide stability and then to insert the register pins into the holes (Fig. 3.8). The subsequent job components are then registered over the first and the adhesive tabs are placed over the pins and attached firmly to the material (Fig. 3.9).

On large jobs where the tendency for movement increases and registration becomes more critical, stick-on registration tabs are often added to the tail (side opposite to the gripper edge) and/or the sides of the reproduction components to provide added stability.

For the system to be effective, it is necessary to have two sizes of pins; low pins for use in vacuum printing frames or to accommodate two or three overlay components and high pins on which three or more components can be overlaid (Fig. 3.10).

Fig. 3.10 Low and high pins

CHAPTER FOUR

Automatic plotting and output

4.1 Impact printers

Conventional impact printing devices (Fig. 4.1), normally associated with the output of textural information, can also be used for plotting rudimentary maps (Fig. 4.2).

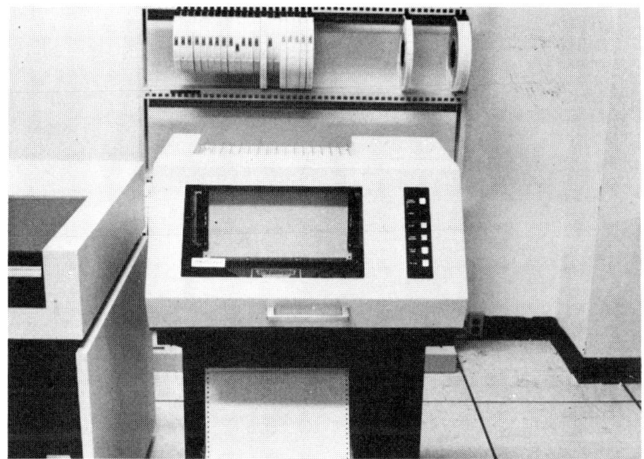

Fig. 4.1 Impact printer

In order to use the printer for this purpose, the system must be capable of breaking up the cartographic data, be it in vector or raster format, into a matrix which corresponds to individual character positions along a line of the printer. Characters must then be inserted into the proper positions in the matrix so that the printed data represents the geographic data upon output.

Different alphanumeric characters can be used to give the effect of different line weights and shading. Also, characters can be double struck or different characters can be overprinted resulting in different densities.

Any output system of this type will have limited resolution (one character space) and accuracy. This system has a limited output size and is restricted to the width or length of the paper.

More recently, dot matrix printers with graphics capabilities have become popular. Individual characters and graphics are formed by a matrix of pins, and this class of printer offers much improved graphics resolution when compared to conventional line (impact) printers.

Dot matrix printers are also restricted to small paper format. Except for the production of simple maps to illustrate reports, impact printing devices are seldom, if ever, used in the production of maps where quality and accuracy are important.

4.2 Flatbed plotters (mechanical/optical)

A class of computer-controlled plotting devices, normally referred to as flatbed plotters, is among those most commonly used for the output of high quality cartographic data, suitable for producing lithographic reproduction materials.

The name 'flatbed' is derived from the plotting surface employed: a flat table or bed to which the plot medium (e.g. paper or film) is affixed (Fig. 4.3). In operation, the medium is usually held in place with adhesive tape; however, some plotters utilize a vacuum system which holds the plotting medium in place. The maximum plot size is constrained by the size of the plotting surface which can range in size from approximately 22 × 28 cm (8.5 × 11 in) to approximately 152 × 152 cm (60 × 60 in).

Fig. 4.3 Flatbed plotter

Flatbed plotters can employ various types of plotting tools and media, depending upon what type of output is required. Plotting tools include pencils, ball point, felt tip or liquid ink pens, scribing points (Fig. 4.4) and cutting knives. Plot-

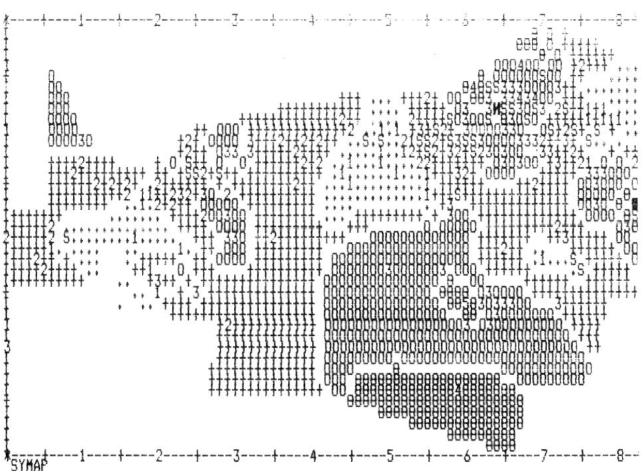

Fig. 4.2 SYMAP output on a dot matrix printer

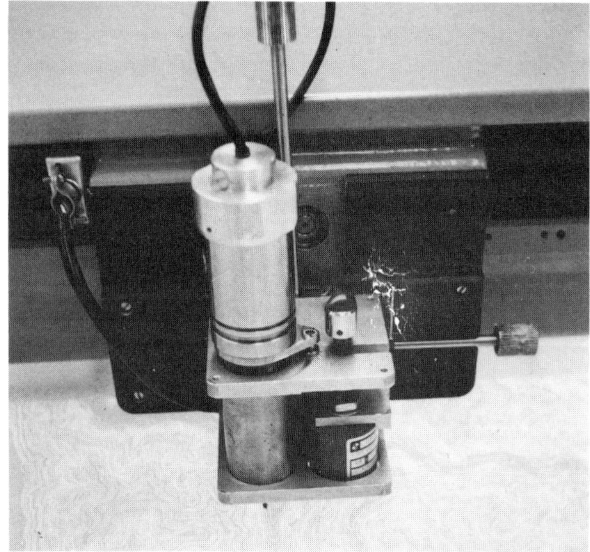

Fig. 4.4 Flatbed plotter with scribing attachment

ting media includes paper, film, scribecoat and masking film.

Flatbed plotters can also be used optically. The plotter is equipped with a light emitting photo head which moves over photographically sensitive plotting material, and depending upon the type of film used, positive or negative images can be produced. Photo heads can be used to generate line work or to produce point symbols. Varying line widths are generated by passing the emitted light through different sizes of apertures or slits in the photo head.

The accuracy of flatbed plotters can be measured in terms of the resolution (the smallest movement the plotting head is able to make) and the repeatability (the accuracy with which the plotting head can return to a given location); both of these measures are usually in the order of plus or minus several tenths of a micron.

Different plotters operate at different speeds. Additionally, the plotting speed may have to be adjusted depending upon the types of tools and media in use. Plotting speeds, characteristically, vary between several tenths of mm/sec to several hundred mm/sec.

4.3 Drum plotters

A widely used class of computer-controlled plotting devices is the drum plotter, which is commonly used for the output of single or multicoloured proof plots.

The plotting media, either paper or drafting film, is affixed to a large sprocketed cylinder or drum

which rotates in the Y direction, and can move in both directions. A plotting pen is mounted on a stationary arm parallel to the axis of the drum. The pen holder moves back and forth across the arm, in the X direction, at right angles to the direction of the drum's rotation across the full width of the drum (Fig. 4.5).

Fig. 4.5 Drum plotter

The width of a plot is restricted by the width of the drum which may be up to 127 cm (50 in). Since the paper or film is fed into the plotter continuously from a large roll, the length of a plot can be much larger than the width.

The colour of plot may be changed by stopping the plotter at predetermined points (e.g. change in features such as roads to contours) and physically changing the pen with a pen of another colour. Newer models have a cartridge which carries several pens of different colours and changing pens is controlled through software commands.

Drum plotters generally have resolutions of 0.25 mm (11/100 in) or better. All movements of the pen are incremental, either in the X direction, or the Y direction. All currently available drum plotters allow simultaneous X and Y movements thus facilitating the smooth plotting of 'off-axis' line work.

4.4 Laser output devices

Advanced laser technology has led to the development of a relatively new class of high precision plotting devices suitable for plotting digital cartographic data. Laser printers and plotters offer flexibility similar to electrostatic plotters, coupled with high resolution normally associated with flatbed plotters.

Fig. 4.6 Laser printer

Laser output devices are divided into two main classes: laser printers and laser plotters. Laser printers (Fig. 4.6) produce high quality output on paper, by using a laser to impart an electrostatic charge to the paper, which is then passed under a toner (pigment) that adheres to the charged image portions of the paper. The image resolution of laser printers is in the order of 120 dots per cm (300 dots per in) and the paper used is approximately 22 × 28 cm (8.5 × 11 in) or 22 × 36 (8.5 × 14 in).

Laser plotters (Fig. 4.7) utilize a laser beam to expose an image onto a photographically sensitive paper or film to produce positive prints and film positives or negatives. This makes laser plotters ideal for directly producing lithographic films ready for plating and printing maps. The plotting media is usually attached to a drum which rotates at high speed. Operating in a raster mode, the laser moves across the surface of the media exposing the image line by line. When plotting is complete, the paper or film is removed from the plotter and the image is developed in a normal manner. The diameter

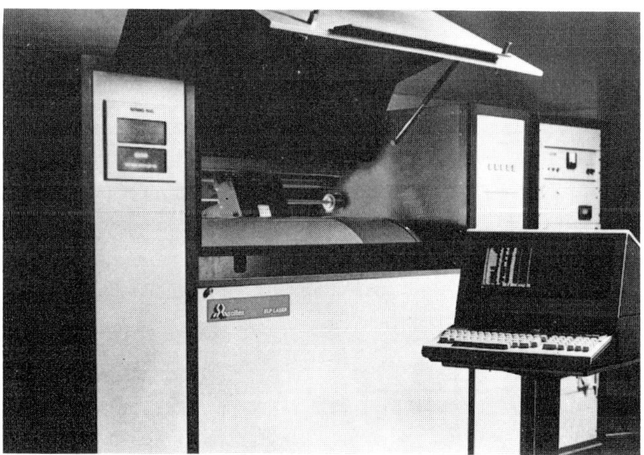

Fig. 4.7 Laser plotter

and width of the drum dictates the maximum size of the plotting media which can be used, which is usually in the order of 101 × 101 cm (40 x 40 in).

Laser plotters usually offer variable plotting resolutions. Commonly, resolutions range from 120 dots per cm (300 dots per in), to 1 200 dots per cm (3 000 dots per in). Since laser plotters use photosensitive materials, a clean darkroom environment is essential.

The high resolution of the laser plotter generates high quality cartographic linework, point symbols and text. Additionally, the laser plotter can be used to produce custom size dot screens for contact printing (see 7.2). The laser plotter can produce combined negatives and positives of line work and screened areas in one pass through the plotter.

Many laser plotters can also be used as a raster scanning (data capture) device, by mounting the appropriate scanning head attachment on the plotter, allowing the automatic digitization of cartographic data.

4.5 Electrostatic plotters

Electrostatic plotters are raster plotting devices which utilize specially prepared paper or drafting film as an output media (Fig. 4.8).

Because the electrostatic plotter is raster based, it plots images in a pixel format, and any data which is captured or stored in vector format must first be converted to raster format before being plotted.

Electrostatic plotting of digital data begins by passing the plotting media over an electrically charged bar which spans the width of the plotter. The bar imparts an electrical charge onto the plotting media (i.e. paper or film which is chemically treated to make it electrically sensitive). This bar is electrically charged with successive lines of the raster image, and imparts the charge to the plotting media as it moves past the bar. Consequently, via electrostatic transfer of the charge, the plotting media becomes charged with the raster pattern of the image. The plotting media then passes through a bath containing the printing chemicals. These chemicals are attracted to the electrically charged areas on the media to form the image. It then passes through heated rollers which fix the image to the plotting media.

Single and multicoloured electrostatic plotters are available. Multicoloured plotters usually require the plotting media to pass through various colour pigment baths, one for each primary colour:

Fig. 4.8 Electrostatic plotter

red, yellow and blue. Other colours are obtained by combining the three primary colours in varying densities.

Because electrostatic plotters are raster based, actual plotting times are independent of the amount of data being plotted, which may be a beneficial consideration for an application where time is a critical factor. Most electrostatic plotters are of medium resolution and are ideal for making proofs of plots of digital cartographic data.

CHAPTER FIVE

Contact copying at scale

Contact copying is both a photographic and photomechanical operation used to produce same size line, half-tone and continuous-tone positives and negatives by a direct contact process. It is also used to prepare drawing or scribing guides in various forms, the production of peel coats for masks for colour tints and to prepare colour proofs and printing plates.

Fig. 5.1 Assembling materials in a vacuum printing frame

To maintain the line widths and other details of the original document, the light-sensitive emulsion side of the material on which the copy is to be made must be brought into direct contact with the original. This is done in a contact or vacuum printing frame which holds the assembled materials in perfect contact when the air has been exhausted from the frame to create a vacuum (Fig. 5.1). The frame is then tilted from its normal horizontal position to a vertical position, so that the glass faces the exposing light located at a distance far enough away so as to ensure even illumination over the whole area of the frame (Fig. 5.2). If the frame is of the flip-top variety, then it is inverted

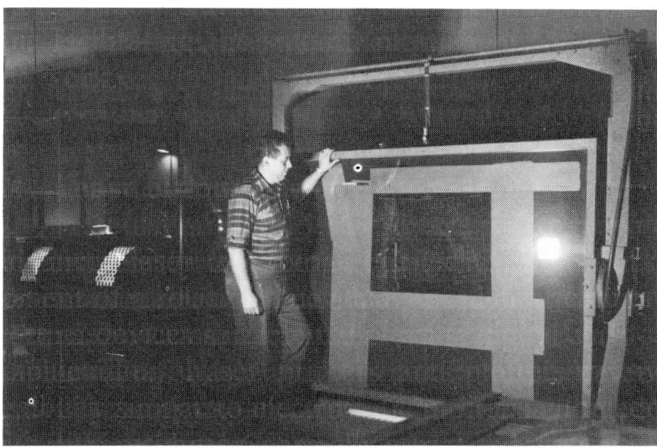

Fig. 5.2 Exposing assembled materials to light source

and exposed to the light source which is underneath.

In the following descriptions, the symbology used for the diagrams illustrating the copying arrangements and flow diagrams of reproduction sequences is shown in Fig. 5.3.

5.1 Photographic contact copying of line negatives and positives

Photographic contact printing is used when there is no change of scale involved, such as in the making of negatives from transparent positive originals and vice versa. It is also used to combine two or more different negatives to make a single positive component which can subsequently be contacted to negative form.

There are many different types of photographic materials used in contact printing. With the exception of a few special low sensitive contact materials, such as bright light, auto positive and duplicating film which can be handled in subdued room light, most other photographic materials must be handled in a darkroom equipped with the usual facilities for processing.

Some of the line copy elements from which contact positives or negatives can be made include originals drawn with ink on transparent and translucent stable base plastics, scribed line work, transparent type and symbol overlays. If a direct positive contact material, such as auto positive film, is used, then a negative can be made directly from a negative and conversely, a positive can be made from a positive.

To ensure high quality reproduction of lines, it is essential that all positive line details are sharp and opaque on a clean transparent or translucent base material and that all negative lines are sharp and clean on an actinically opaque background.

In contact printing, it is important to know whether the original negative or positive is right-reading or wrong-reading as this will determine the type and quality of copy required. The term right-reading is applied to a document on which the lettering reads normal from left to right whether in positive or negative form when viewed from the emulsion side of the film (Fig. 5.4). The term wrong-reading is applied when the lettering reads from right to left whether it is in positive or negative form when viewed from the emulsion side of the material (Fig. 5.5).

Normally, contact copies are made with the emulsion side of the original placed in direct contact with the emulsion side of the material on which

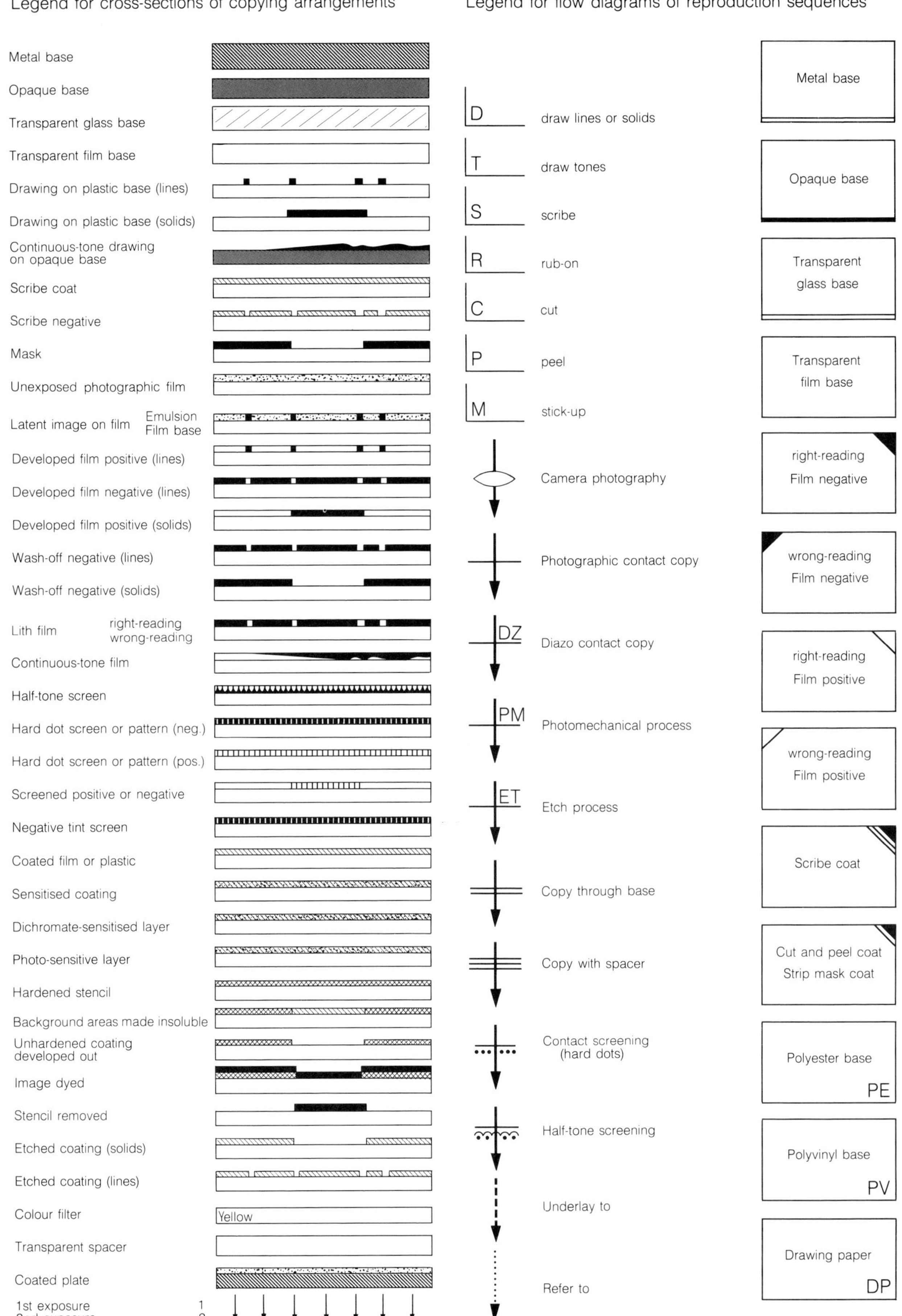

Fig. 5.3 Legend for cross-sections of copying arrangements and legend for flow diagrams of reproduction sequences

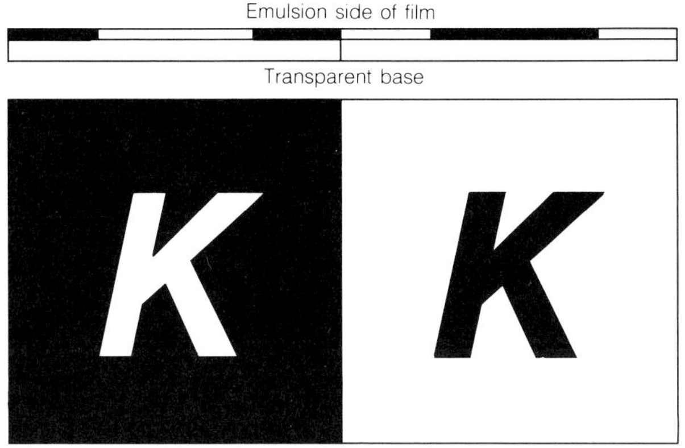

Fig. 5.4 Right-reading negative and positive viewed from the emulsion side of film

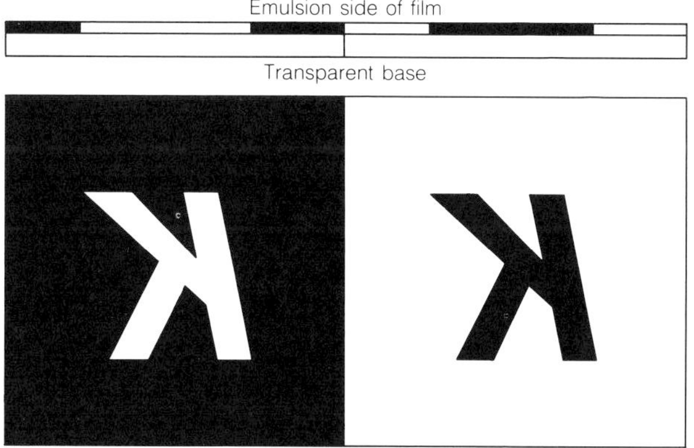

Fig. 5.5 Wrong-reading negative and positive viewed from the emulsion side of film

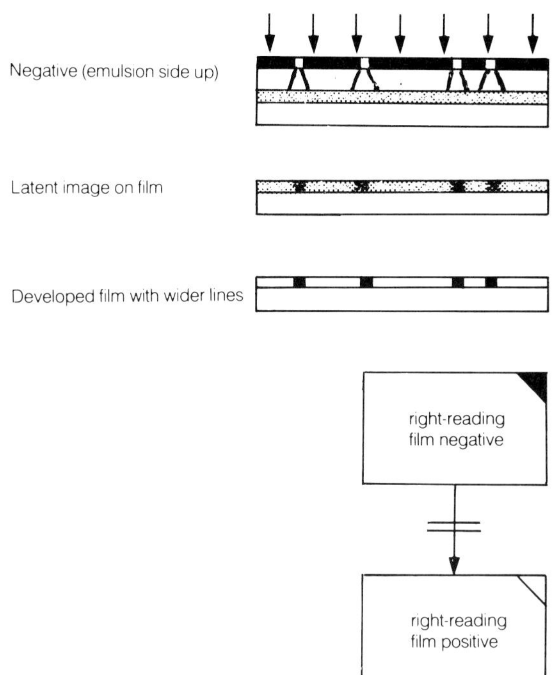

Fig. 5.7 Contact copying through base of negative to produce wider lines

the copy is to be made (Fig. 5.6). However, if a copy with wider lines than those of the original is required, it can be achieved by copying through the base of the negative with the emulsion side of the film up (Fig. 5.7); this allows the light to spread slightly between the materials during exposure. When the exposed film is developed, the lines will be slightly wider in positive form. Similarly, a positive can be used to produce thinner lines in negative form.

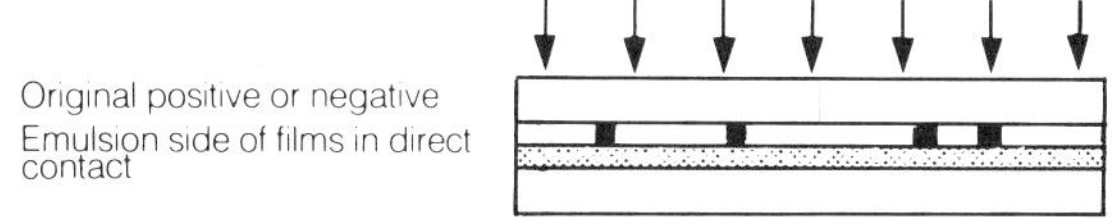

Fig. 5.6 Normal photographic contact copying with the emulsion side of original positive or negative in direct contact with film on which copy is made

5.1.1 Negative process in contact copying at scale

Negative contact copying at scale is the most common and widely used method to reproduce lines, solids and tints from original positive documents.

The lines and solids on the original document must be sharp-edged and totally opaque on a clean transparent stable base material. Where line rulings, dot patterns, symbols and lettering have been applied on the original, they must also be opaque and lie perfectly flat on the surface of the document.

The dimensional stability of both the originals and the film bases is of the utmost importance in cartography. Therefore, polyester based plastics and films, because of their high dimensional stability, are mainly used in the process.

A negative contact copy is made by placing the original positive image in direct contact with the light sensitive side of the film in a vacuum printing frame. The air is then exhausted from the frame and the assembled material is exposed. During exposure, the light passes through the transparent parts of the original (Fig. 5.8) and forms a latent image on the emulsion of the film.

Conversely, the light is held back by the actinically opaque parts of the original image. Following exposure, the latent image on the film is converted into a permanent visible image when developed in a chemical solution that reduces the expanded silver halide in the emulsion to a black metallic silver to produce a negative of the original document. The type of film and developer used in the process is normally of the lith or line variety. This combination of film and developer allows for

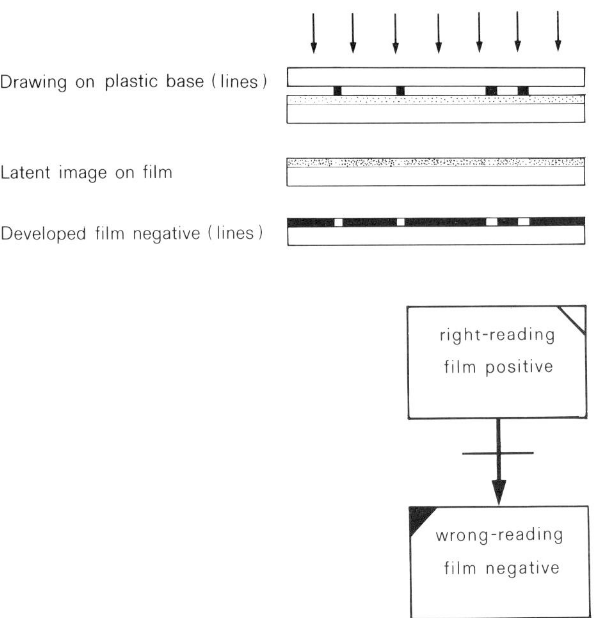

Fig. 5.8 Contacting arrangement for a positive to negative copy

optimum reproductions that accurately preserve the line widths and details of the original image. The density profile of the copy (Fig. 5.9) shows the step flanks that are due to a high gamma of the characteristic density curves for lith and line films and developers (Fig. 5.10).

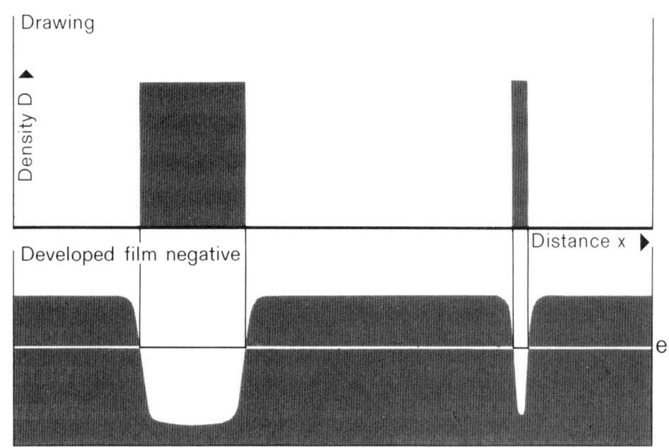

Fig. 5.9 Density profile of drawing and resulting negative

For an accurate further copy from this negative, care must be taken to expose and develop the film to level 'e' in the diagram shown in Fig. 5.9. Any over or underexposure would result in slightly thicker or thinner lines.

The process is also used to produce positives from negative originals.

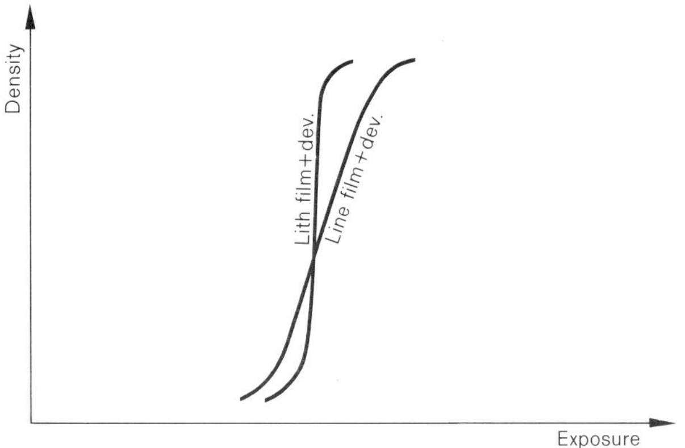

Fig. 5.10 Characteristic density curves for lith and line films and developers

5.1.2   *Direct positive process*

Direct positive film is used to produce positive copies from positive originals, negatives from negative originals and reflex copies from opaque originals by contact copying. The contacting is done in a normal vacuum printing frame in which the original document and film are exposed using a very high intensity light source.

The type of film that is used has been pre-exposed during its manufacture to its maximum latent density, so that additional exposures to the emulsion will reduce the latent density. Simply stated: if the unexposed film were to be developed, an overall even black image would result. If however, the material is pre-exposed through a yellow filter, it will be rendered a negative working film and will produce a negative from a positive and, conversely, a positive from a negative when exposed to a high intensity light source.

It is possible to produce a combined positive from a combination of one positive and several negatives. This is achieved by first exposing the positive through a yellow filter to produce a latent positive image and rendering the unexposed areas of the film negative working. The multiple negatives can then be exposed in turn without the yellow filter. The film when developed will contain a combined positive image.

Figure 5.11 illustrates the copying arrangement for a positive to positive copy. The built-in latent density disappears in the area exposed to the yellow light resulting in a reverse, or wrong-reading positive when the film is developed in the normal manner.

Reflex copies (see 5.1.4) can also be made from ink lines drawn on an opaque surface, such as paper, by placing the direct positive film, with the emulsion side in contact with the image side of the

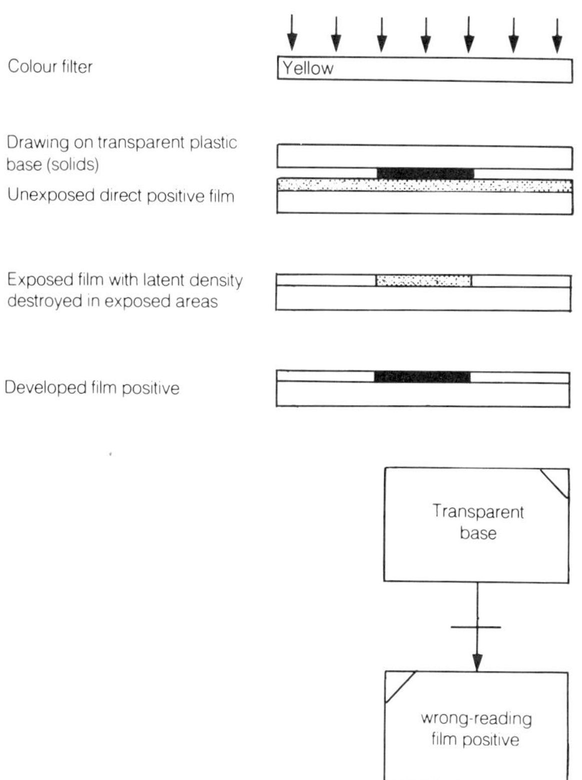

Fig. 5.11 Contact copying arrangement for a positive to positive copy using direct positive film

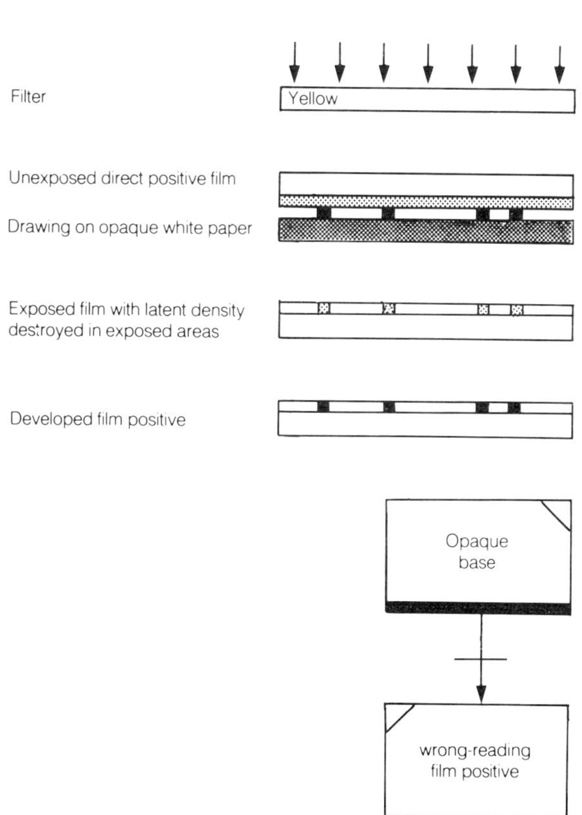

Fig. 5.12 Exposure arrangement to obtain a reflex copy of an ink drawing on opaque white paper

paper, in a contact printing frame and exposing with a yellow light through the film base (Fig. 5.12).

Other uses of direct positive film include the production of lateral reversals, i.e. changing the image from right-reading to reverse or wrong-reading on the emulsion side of the film. It is also used to duplicate other map reproduction components to preserve them in separate storage and for use in the production of other derived maps and similar documents.

5.1.3 Wash-off process

Polyester base wash-off films are characterized by a different light-sensitive photographic emulsion layer which, after exposure, does not have to be developed in the normal manner.

The original negative and wash-off films are normally assembled in a vacuum printing frame and exposed using a high intensity light source. Following exposure, the light hardened latent image is developed with an activating solution and the unexposed parts of the film are simply rinsed away with water (Fig. 5.13). It is not necessary to chemically fix the activated image.

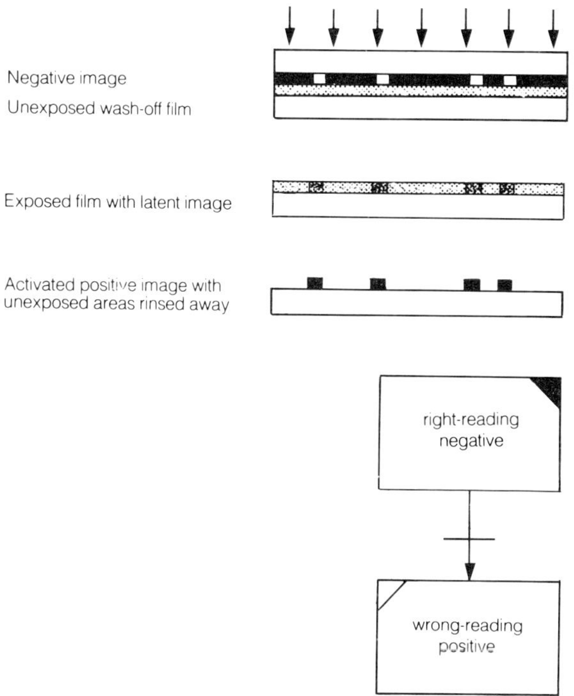

Fig. 5.13 Exposure arrangement for wash-off film

Wash-off films can be obtained as contact films, direct positive films, and projection films. The first two types can be processed under subdued daylight conditions.

Wash-off type film can be useful for revision purposes as incorrect or obsolete line details can be erased with a wet vinyl rubber, and new, or additional line detail, can be added directly to the surface of the film with a pen and an ink suitable for drawing on polyester base materials. The revised positive can, of course, be contacted to negative form if required.

5.1.4 Reflex process

The reflex method of contact copying is a cheap and simple process by which copies of opaque originals can be made without resorting to the use of a camera. As the name implies, the technology is based on the reflection of light rays from the surface of the original back to the emulsion surface of the film.

In preparing to make a reflex copy, the opaque original (e.g. an ink drawing on paper) to be copied is placed in a vacuum printing frame with the image in direct contact with the emulsion side of the reflex film. When exposed through the base of the film, the light rays pass through the film to the surface of the original copy. The light, in turn, is reflected from the white areas of the original back to the emulsion side of the film, whereas, the light hitting the black image of the original is absorbed (Fig 5.14). The result, after developing, is a mirror reversed image of the original. Using this process, a positive copy can be made from a positive original and a negative copy from a negative original. Direct positive film (see 5.1.2) can also be used to make reflex copies.

Reflex film basically consists of three layers: an upper layer with a very thin silver diffusion coat, a thicker middle layer of silver bromide and a third layer which serves as a colour filter on the back of the film. The function of the filter layer is to let the light rays of certain wavelengths pass through the film to the original. If an ultraviolet light source is used to expose the film, an extra yellow filter is placed between the film and the light source to achieve the desired wavelength.

In general, polyester base reflex films, because of their very good dimensional stability, are well suited for cartographic purposes. They are available in two versions: glossy and matt surfaced. The glossy film has the advantage of having better printing quality, however, additional detail may be drawn with water-soluble inks on the surface of the matt films.

5.2 Photographic contact copying in continuous-tone

Continuous-tone positive or negative copies of original images have a complete range of tones, from the lightest highlight areas of the original to the darkest shadow areas. Copies of these negatives, or positives, are made in a vacuum printing frame either directly onto film or a photographic paper.

The continuous-tone process has limited use in cartography. Its most frequent application is in the making of paper prints from rectified aerial photographic negatives for use in the production of photomaps or for other special requirements, such as, in the reproduction of shaded relief drawings.

Before a continuous-tone image can be reproduced on a printing press, the image must be broken down with a half-tone dot screen, by photography or by contact copying.

High contrast positives and negatives have the greatest tonal range between the darkest and lightest areas of the original image, whereas, low contrast positives and negatives have little separation between the lightest and darkest tones. However, overall contrast can be reduced or increased during processing so as to match adjoining prints used, for example, in the construction of mosaics and photomaps.

In the contacting process, it is helpful to have a

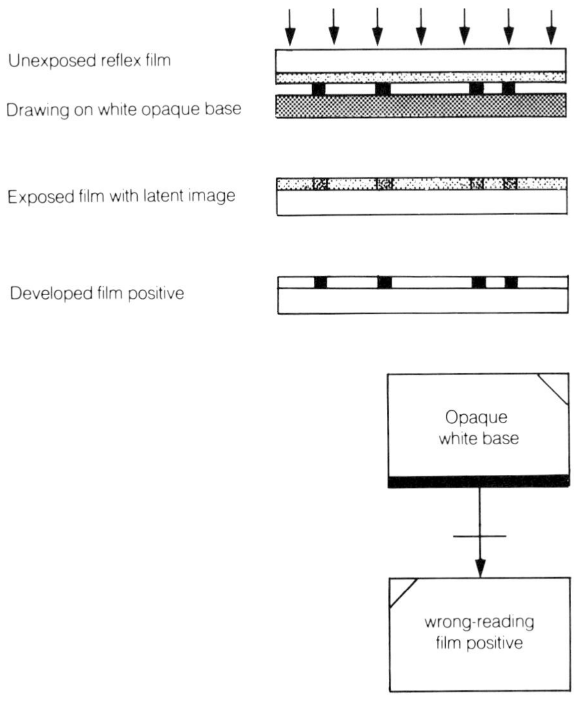

Fig. 5.14 Contact copying arrangement for reflex copies

continuous-tone grey scale with a series of grey tones ranging from white to black attached to the edge of the original positive or negative, as it provides a visual indication of the separation between the darkest, middle and lightest tones on the contact copy. For exact control, a densitometer should be used to measure the densities of continuous-tone originals prior to contact printing or photography in order to achieve the best required results. If under-exposed, the gradation in the highlight areas of the image will be poor and the black shadow areas will be grey and not black enough. In the case of over-exposure, the highlight

and a smooth paper will produce low contrast prints.

5.3 Photographic contact printing of continuous-tone in colour

Contact printing of continuous-tone positives or negatives in colour has a limited application in cartography. Its main use is in making photographic contact paper prints from coloured aerial photographic negatives for use in preparing mosaics and photomaps in colour.

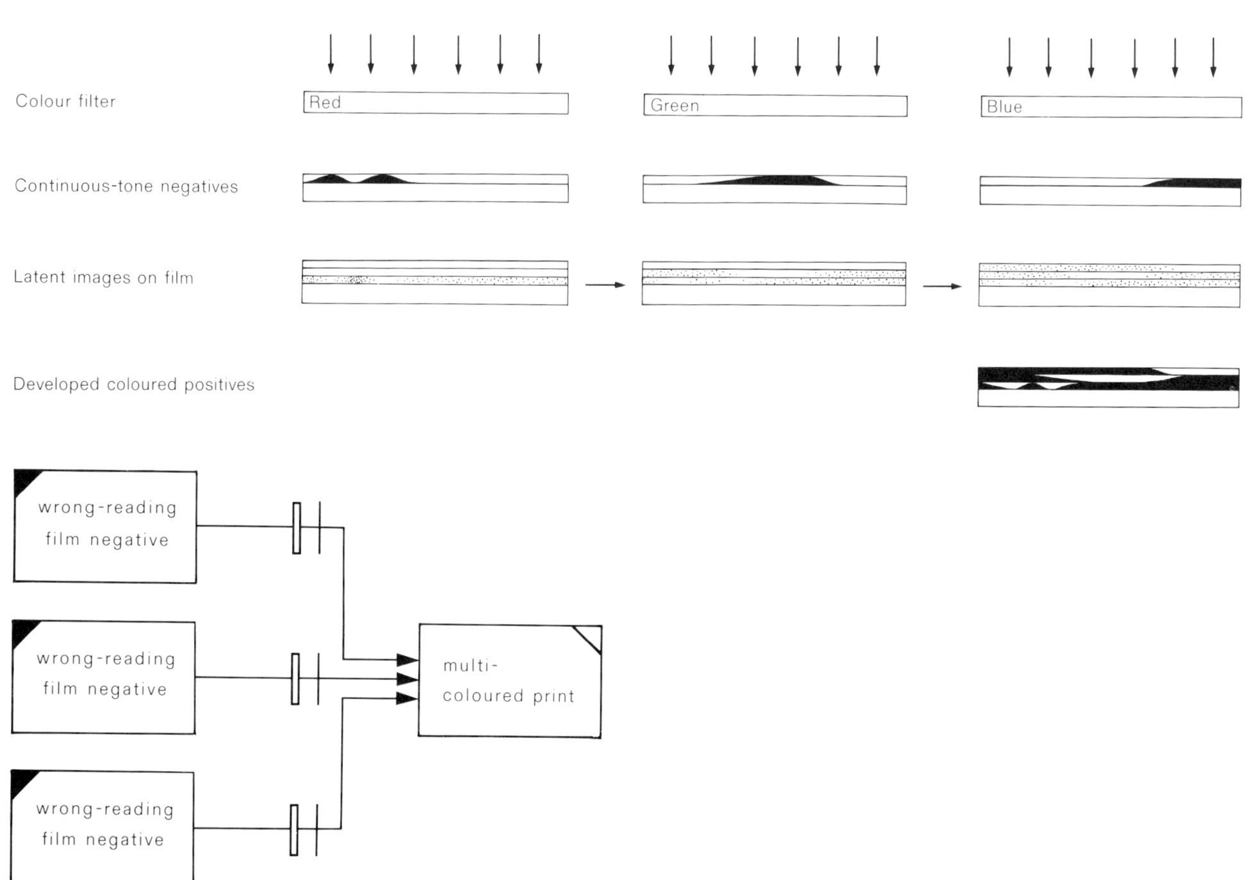

Fig. 5.15 Contacting sequences using separated continuous-tone black and white negatives

areas will be grey and the shadows will not clearly differentiate detail.

There are many different grades of photographic papers that vary in contrast, texture and thickness. They can be characterized as having normal, hard or smooth gradations. Almost any type of negative can be used to make prints when papers with corresponding gradations are used. Generally, a normal negative copied onto paper with a hard gradation will produce high contrast prints

Coloured continuous-tone copies can also be made with the use of three colour separated continuous-tone black and white negatives, by exposing them sequentially with red, green and blue light onto a photographic colour printing paper (Fig. 5.15). These colour prints can be used as colour proofs for editing purposes, or they can be used as original copies for half-tone colour separation by photography (see 7.1.1 and 8.3) and subsequent process printing.

5.4 Contact printing with dichromated colloids

Contact printing using dichromated colloid coatings is extensively used in the cartographic reproduction processes. Its principal uses are in the making of guideline images for drawing and scribing, colour proofing, positive to positive image duplication, tint and line combinations and the making of etched images on strip mask material for subsequent peeling of area masks. Dichromated colloids are also used to transfer images to printing plates for lithographic offset printing (see 10.1); silk screen printing (see 10.4); letterpress printing (see 10.2); photogravure (see 10.3) and collotype printing (see 10.5).

The sensitizing solution for dichromated colloid coatings consists of an organic or synthetic colloid disolved in a solvent, usually water, to which a soluble dichromate is added. The coating is applied to the surface of the material onto which the image is to be transferred by spreading it evenly with a whirler or some other type of coating device.

Copies can be produced from both negative and positive originals. In either case, the original document is placed in direct contact with the coated surface of the material in a vacuum printing frame and exposed to a high intensity ultraviolet light source. If a negative is used for exposure, the unexposed non-image areas are removed in the developing process and the image is formed by applying a dye to exposed light hardened image

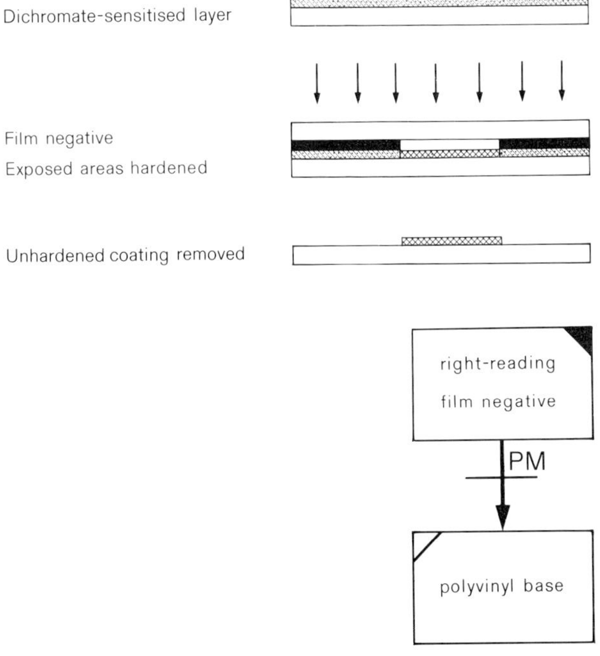

Fig. 5.16 Exposure and developing sequences when a negative is used

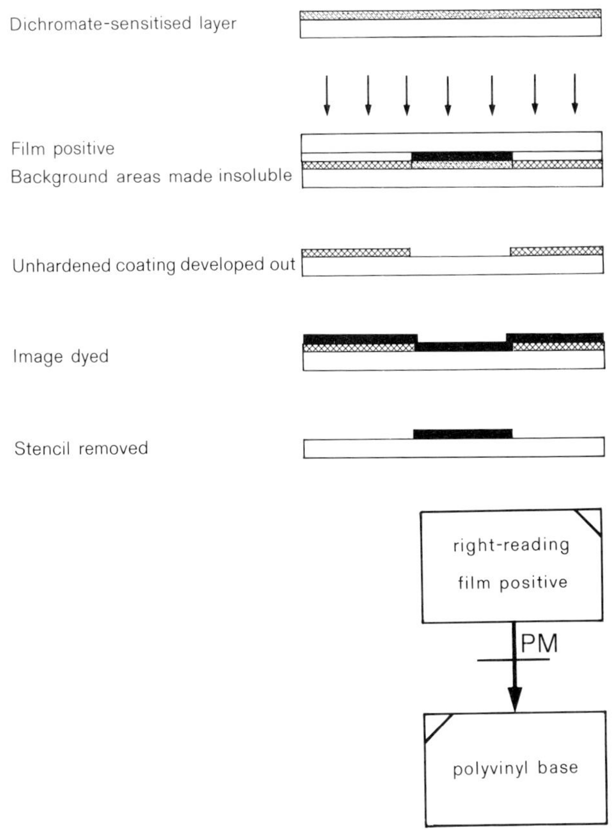

Fig. 5.17 Exposure and developing sequences when a positive is used

areas (Fig. 5.16). If a positive is exposed, the unexposed areas are removed in processing with an aciduous solvent, leaving a resistant light hardened stencil surrounding the image areas. A coloured deep-etch lacquer is then applied over the stencil. The stencil is removed by rinsing with water which carries the surplus lacquer with it leaving behind a positive image on the supporting base (Fig. 5.17). The construction of an image from more than one positive involves re-coating the supporting material as the whole process must be repeated for each image element. This is necessary when two or multicoloured images are required.

If plastic sheets are used, they can be transparent, translucent or opaque and based on polyesters, polycarbonates and vinyl polymers. The latter type plastics have the advantage, especially in the positive process, as they absorb the dye permantly into the base material, whereas, the polyesters have to be pre-coated in order to accept the dyes.

5.4.1 Contact printing with photopolymer film

Photopolymer film consists of a light-sensitive photopolymer layer supported on a polyester base material with an antihalo layer to prevent the reflexion of light from the reverse side of the film.

The contact copying process, using this type of film, is similar to contact printing using dichromated colloids (see 5.4). A positive original is placed in direct contact with the light-sensitive photopolymer layer of the film and exposed in a vacuum printing frame using an ultraviolet light source. Following exposure, the unexposed soluble image areas are washed away from the sensitive layer with a weak soda solution (the antihalo backing is also removed at the same time) leaving only the light hardened non-soluble areas in the form of a stencil. The image is developed by applying a dye that adheres to the light hardened non-soluble areas to produce a negative image of the positive original (Fig. 5.18).

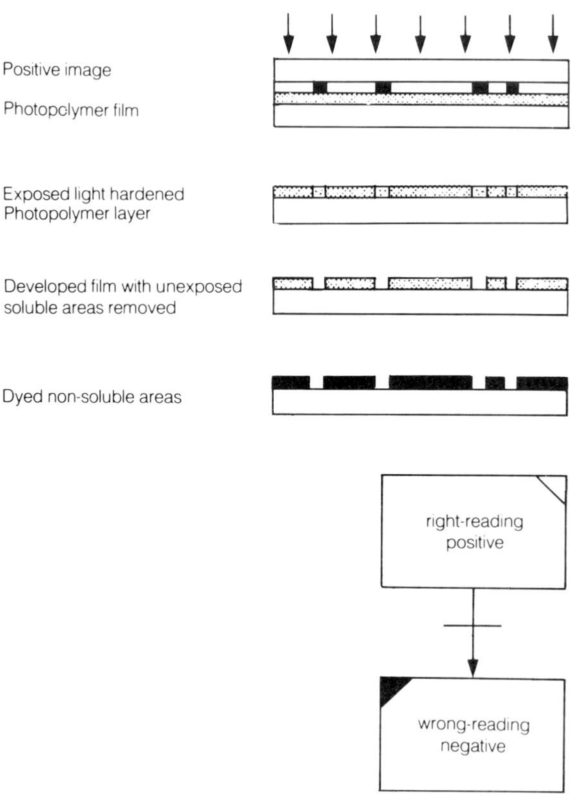

Fig. 5.18 Exposure and developing sequences for photopolymer film

Various types of film are available. Some A4 size films have a built-in adhesive substance in the sensitive layer which, after copying, will permit the image to be transferred to another material by a rub-on-technique (see 1.5).

Photopolymer coatings are also used for producing printing images on normal lithographic printing plates and plates for waterless offset printing.

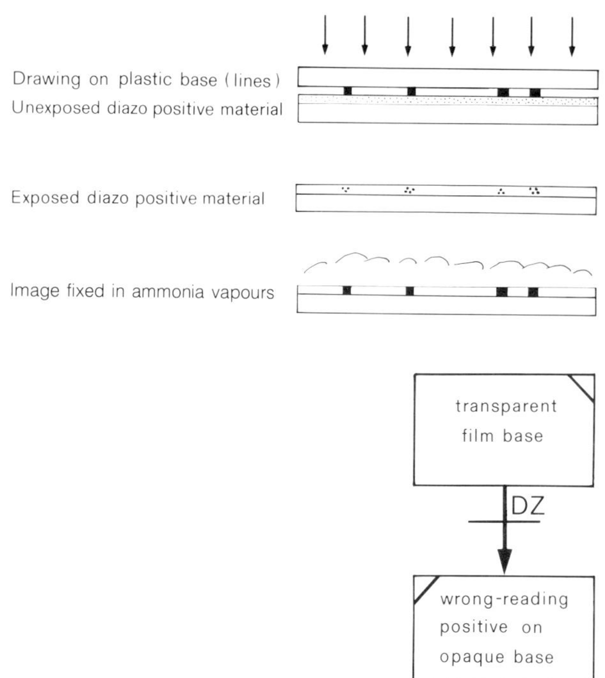

Fig. 5.19 Positive to positive copying process

5.5 Diazo contact copying

Diazo copying is used to make individual or multiple copies of an original document directly onto a material that can be coated with a light-sensitive diazo chemical compound. Although it is primarily a positive to positive copying process (Fig. 5.19), there are also coatings that will provide a positive image from a negative original (Fig. 5.20) and

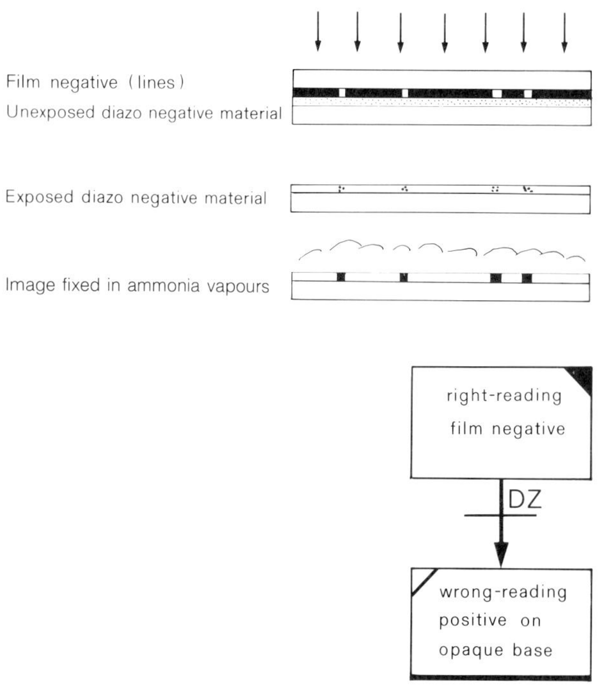

Fig. 5.20 Negative to positive copying process

continuous-tone images from continuous-tone originals.

Diazo coatings can be applied in the form of a solution onto almost any type of base material. Precoated materials, such as rolls of paper, tracing paper and plastics are also available commercially.

Copies of originals are usually produced in a standard model exposure machine which has a built-in capability for continuous development (Fig. 5.21). The original document is placed in direct contact with the light-sensitive surface of the material and inserted into the copying machine where it is pressed against a rotating glass cylinder with the source of illumination on the inside. For more precise copies on dimensionally stable base materials, it is advisable that the assembled materials be exposed in a vacuum printing frame.

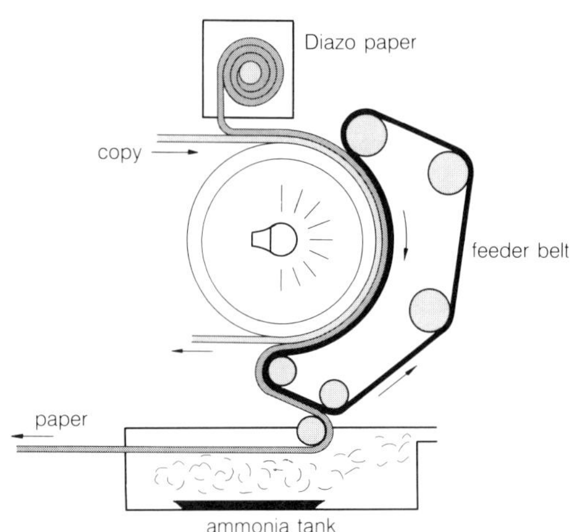

Fig. 5.21 Machine for diazo reproductions

During exposure, the ultraviolet light source destroys the light-sensitive diazo layer in the areas not protected by the original image.

The latent image in the protected areas is developed when the material is exposed to ammonia fumes. (Developing in this case means that the unexposed azo compounds change colour and become light resistant. Normally the result is a black or brown image.)

Diazo copies can also be produced in other colours such as cyan, magenta and yellow. Although diazo images tend to fade due to the influence of ultraviolet light which is present in daylight and most other forms of illumination, they do provide a quick means of obtaining a proof in colour by using commercially available presensitized transparent foils.The foils are exposed, along with the individual original documents, and developed by ammonia fumes to form a colour image on each separate foil. These foils can then be registered one on top of the other to produce a transparent colour proof (see 9.3).

Some other typical applications of diazo contact copying in cartography include: the making of guide images on scribing materials, producing verification copies that need not be dimensionally stable and to produce film positives as intermediate products.

Diazo films, because of their good resolution and low price, are extensively used in cartography today.

5.6 Cyanotype or blueline process

The blueline process is extensively used in cartography, especially to produce blueline images or 'keys' on paper, plastic and scribing materials in order to provide an accurate guide for drawing or scribing. The image is applied in a light blue actinic colour so that it will not reproduce in any subsequent photographic or contact copying operations.

The process is based on the light sensitivity of ferric salts which when exposed to ultraviolet light is reduced to a ferrous salt which, combined with another substance, usually potassium ferricyanide, produces a blue colour image.

The sensitizing solution for coating the materials, can be derived from a number of formulae. The most common is an aqueous solution containing ammonium citrate and potassium ferricyanide.

An image can be produced from either a positive or a negative. A positive will produce a white image on a blue background, whereas, a negative will produce a blue image on a white background, the latter being the more widely used. The blueline coating can be applied either manually by wiping the light-sensitive emulsion over the surface of the material or with the aid of a whirler. Before coating, the material, especially plastics, should be washed and cleaned with a wetting agent to allow the coating to flow evenly over its surface. The coated material, together with the positive or negative original, is exposed in a vacuum printing frame. Following exposure, the copy is washed in water to remove the unreduced salts from the unexposed areas leaving only the exposed nonsoluble ferrous ferricyanide which turns blue during the process. Figure 5.22 illustrates the exposure and developing sequences when a negative is used to produce a blueline image.

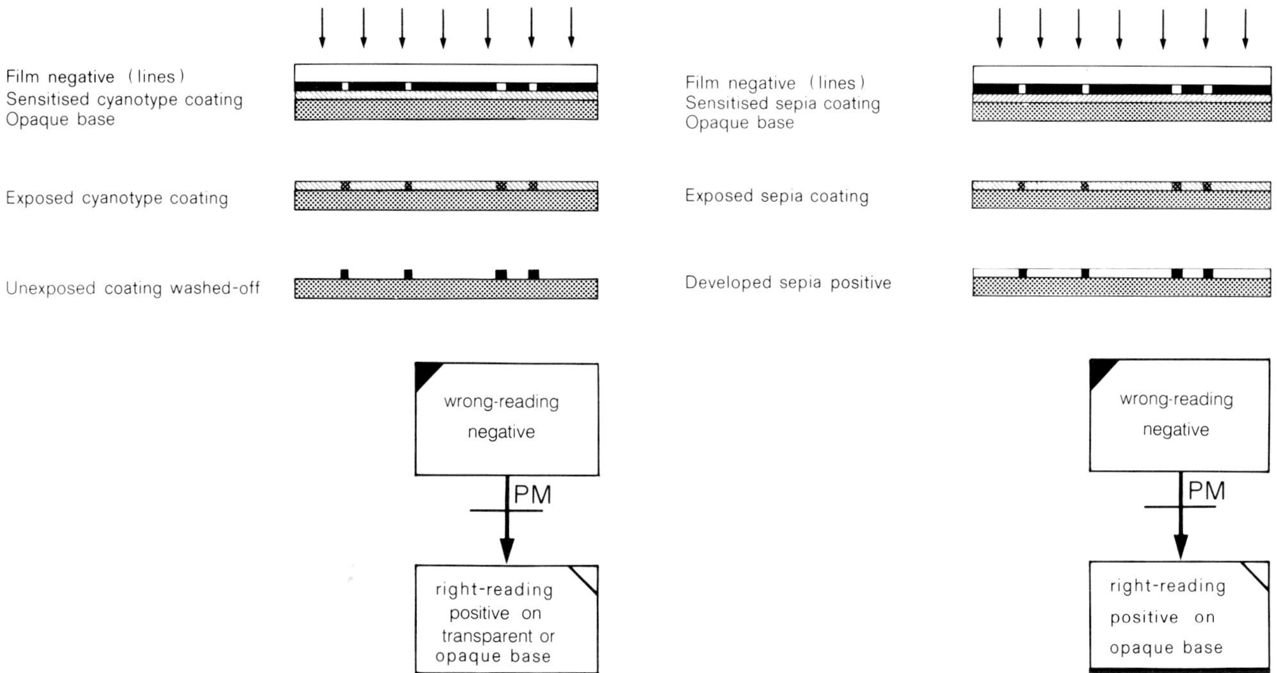

Fig. 5.22 Exposure and developing sequences when a negative is used

Fig. 5.23 Exposure and developing sequences using a negative

5.7 Brown print or Vandyke copying process

A brown print or Vandyke copy is made on an inexpensive photographic paper that has been sensitized with a mixture of iron and silver salts contained in the gelatin or gum coating of the paper. When exposed to an ultraviolet or blue light, the iron or ferric salt is reduced to a ferrous salt which is a potent reducing agent that reduces the silver salt to form a metallic image.

As in the case of the blueline process (see 5.6), brown print copies can be made from either positive or negative originals. A positive will produce white lines on a dark brown background; conversely, if a negative is used, which is nearly always the case in cartography, it will produce dark brown lines on a white background.

Copies are made by exposing the assembled sensitized paper and negative original in a vacuum printing frame. The exposed paper is developed in water which results in a dark brown positive image on the white paper (Fig. 5.23).

It is also possible to make a number of separate exposures onto the same sheet of brown print paper. This is particularly useful in multicolour work as each individual colour negative can be successively exposed in register with each other. The exposure times used are different for each negative. For example, the negative of the black detail is given a full exposure and the time is reduced for each other colour negative. When developed, the results will show a noticeable contrast in the shades of brown for each colour component.

Brown print copying is most commonly used to produce verification copies, checking map content, design etc.

With the introduction of diazo copying methods (see 5.5), the use of brown print copying has declined in cartography. However, it has the advantage of producing a much more stable image in the long term.

5.8 Deep etch scribecoat copying process

It is often necessary to produce images on scribecoat materials in order to add new or revised detail by scribing. This is achieved by the deep etch process, whereby a duplicate image of an original scribed or photographic image is chemically etched into the scribing surface of ordinary scribing materials.

This is accomplished in a vacuum printing frame, in which the emulsion side of the original scribed or photographic negative is placed in contact with the presensitized emulsion side of the scribing material (Fig. 5.24). The assembled material is then exposed to a high intensity light source. During the developing process, the exposed lines

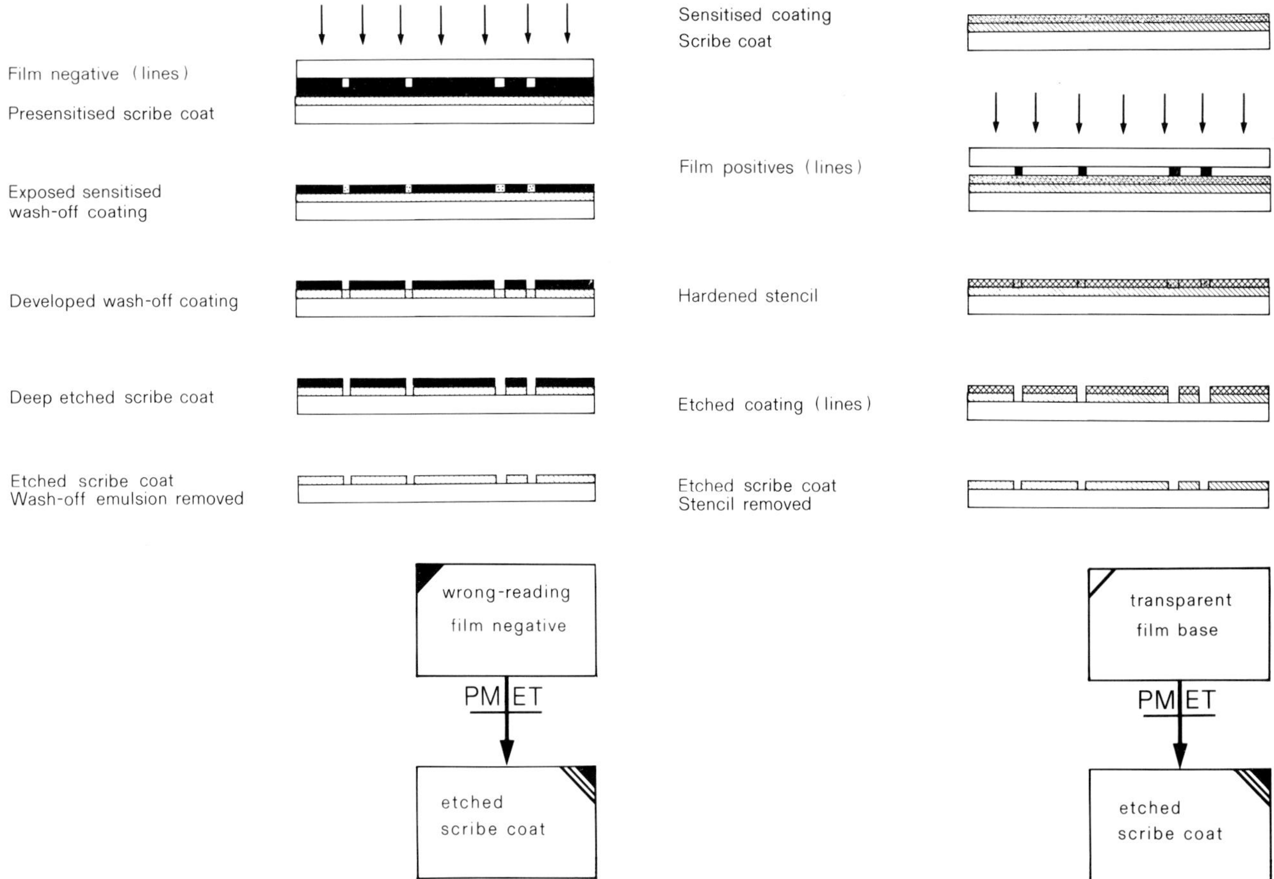

Fig. 5.24 Exposure and developing sequences using presensitised scribing material

Fig. 5.25 Exposure and developing sequences in positive etching process

are removed leaving the unexposed areas in the form of a stencil. A chemical etching solution is then spread evenly over the surface of the stencil and penetrates the scribing layer of the scribecoat through the clear developed areas of the stencil. The emulsion layer on the scribecoat is then washed away carrying the stencil with it, leaving a duplicate negative image of the original negative. The exposure and development sequences for the positive etching processes are illustrated in Fig. 5.25.

Duplicate etched scribecoats are mainly used for revision purposes. For example, obsolete map detail can be deleted from the original scribed or photographic negative by opaquing, after which a duplicate etch scribecoat is made. A blueline guide image (see 5.6.) of the new information, in negative form, can then be added in register with the etched image on the duplicate scribecoat (Fig. 5.26). The new details can then be scribed in the normal manner to produce a revised document.

The etching process must be carefully controlled as lines have a tendency to broaden.

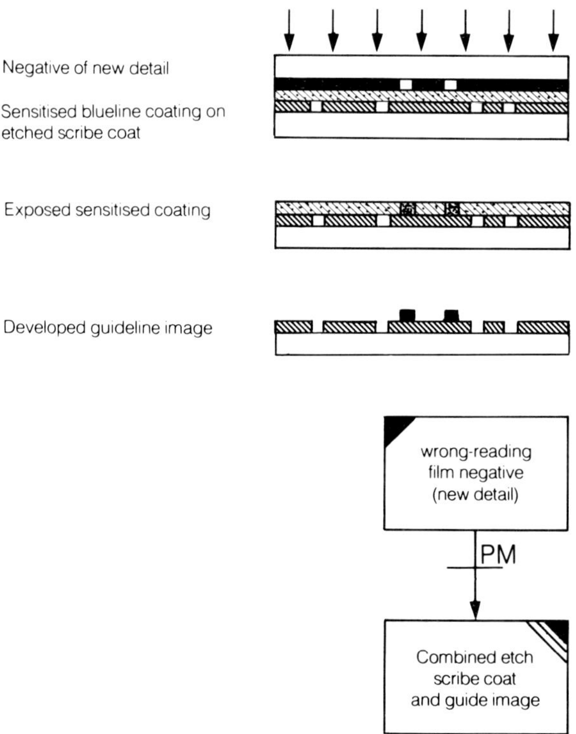

Fig. 5.26 Exposure arrangement for combining a guideline image with an etched scribe coat

5.9 Peel coat or strip mask technique

The peel coat or strip mask technique is used when precise registration is of the utmost importance. It differs from the manual masking techniques described in 1.2.2 and 1.4, as the areas to be masked are derived directly from the outline image of the areas which are to contain tints. There are some commercially available presensitized peel coat materials specially for this purpose that can be used with either negative or positive originals.

If a scribed or photographic negative is used (Fig. 5.27), it is exposed in a vacuum printing frame with the emulsion side of the negative placed in direct contact with the presensitized layer of the peel coat material. During the developing process, the exposed image is removed leaving the unexposed background areas in the form of a stencil. Following this step, the image is chemically etched through the stencil and through the stripping membrane to the surface of its supporting transparent polyester base.

If a positive working presensitized material is used, the exposure and developing sequences are virtually the same, except that the positive original, in the case of some materials, is exposed through the base of the peel coat material. This causes a slight reduction in the line width on the negative mask which contributes to good registration of colours during printing.

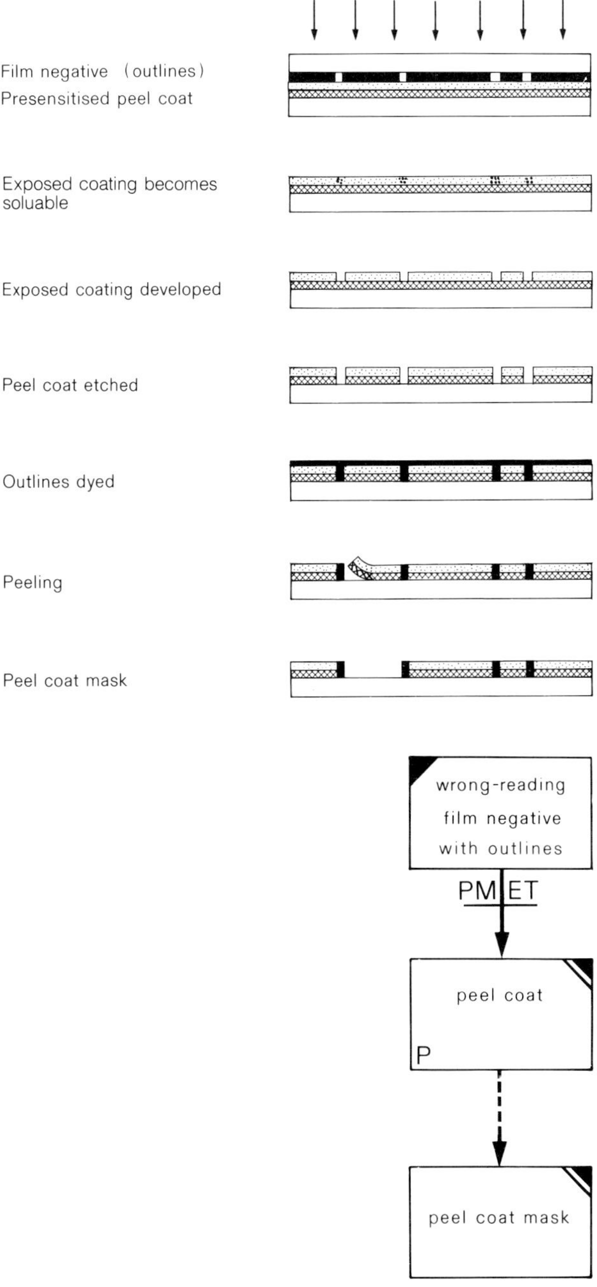

Fig. 5.27 Exposure and developing sequences using presensitised negative working peel coat

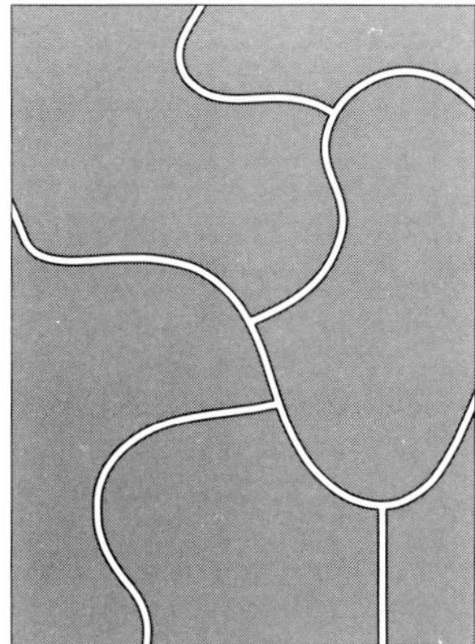

Fig. 5.28 Etched peel coat image

The etched peel coat material will, of course, include all of the details contained on the original negative or positive (Fig 5.28). It is, therefore, necessary to opaque-out the unwanted lines. This can be done by coating the etched surface of the peel coat with an opaque or dye solution to fill in the lines (Fig. 5.29). Once the fluid is thoroughly

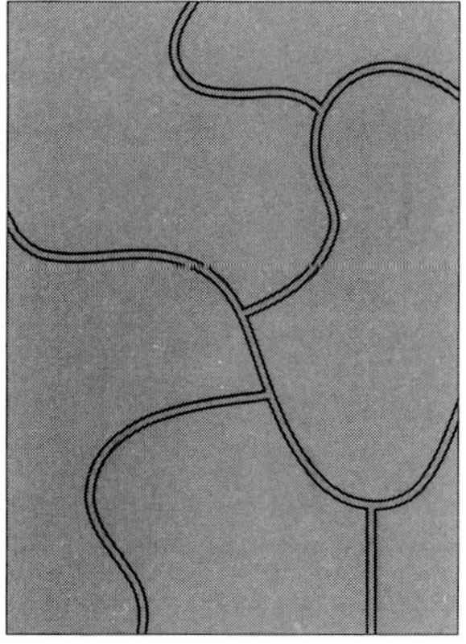

Fig. 5.29 Etched image coated with an opaque solution

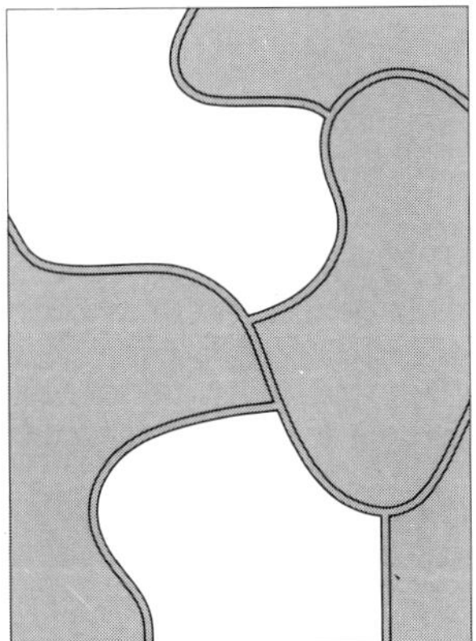

Fig. 5.30 Peeled areas slightly smaller due to opaquing of lines

dry the tint areas can then be peeled away between the lines (Fig. 5.30).

When this system of opaquing is used, it should be kept in mind that the cleared areas of the mask will be slightly smaller due to the opaquing of the lines (Fig. 5.30). However, an excessive exposure at the next stage will usually broaden the area to compensate. If the lines were not opaqued the cleared areas would be slightly larger (Fig. 5.31).

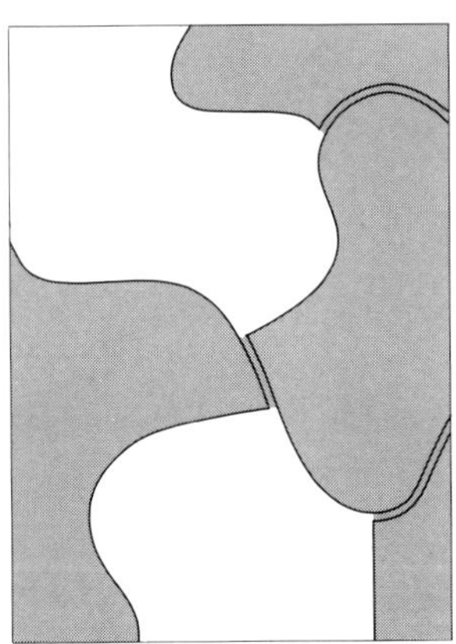

Fig. 5.31 Peeled areas slightly larger when lines are not opaqued

When a map comprises several tint components, a combined negative or positive containing the outline details can be made to serve as a master from which the required number of peel coats can be produced.

For reasons of economy, a single mask (Fig. 5.32) can be used to serve as several masks, for example in the production of hypsometric or bathymetric tints by successively peeling each tint area and contacting them to positive form and finally combining them to produce a single colour component.

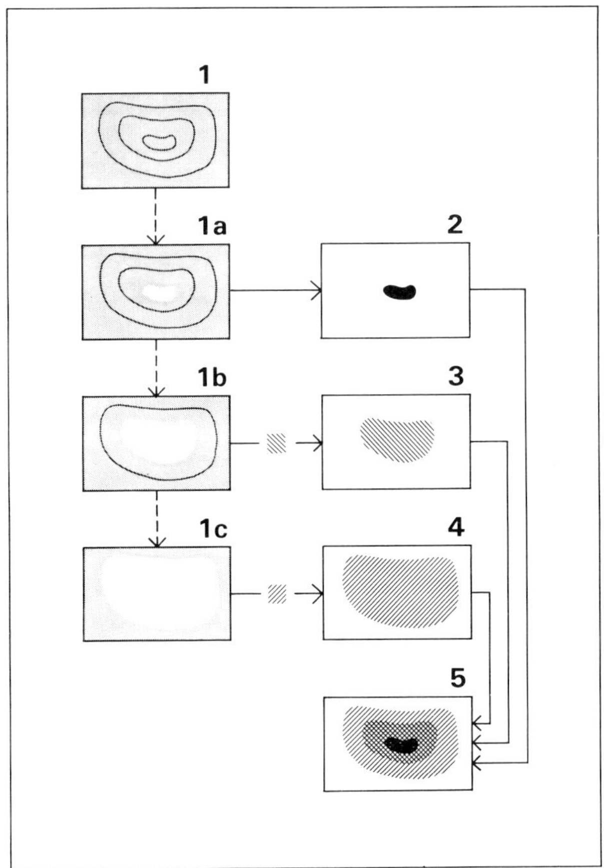

Fig. 5.32 Successive peeling of different areas on the same peel coat materials

5.10 Copying of press plates for lithographic printing

A lithographic printing plate is usually a metal which has been chemically cleaned and coated with a light-sensitive emulsion. This coating is soluble in some solvents such as water, but becomes insoluble when exposed to a high intensity light source. The plate itself consists of two distinct areas, the printing areas which will accept grease and repel water, and the non-printing areas which will accept water but repel grease. In order to accept water, the plate has a grained surface which, if magnified, would appear as a random pattern of peaks and valleys in which the water is retained.

There are basically three types of lithographic plates: surface plates, deep-etch plates and bimetal plates.

Surface plates are available commercially in presensitized form, or they can be coated by hand or in a simple roller coating device with a wipe-on light-sensitive solution. In either case, the exposed light-sensitive coating becomes the printing image on the surface of the plate.

The printing image can be transferred to the

printing plate from either negative or positive originals. In either case, they are placed in direct contact with the coated plate in a vacuum printing frame and exposed to a light source. The light penetrates the transparent areas of the negative or positive and hardens the coating making it insoluble. Where the coated surface is in contact with the opaque areas of the negative or positive, the coating remains soft and soluble.

If a negative is used (Fig. 5.33), following exposure, the printing image is developed by coating the exposed surface of the plate with a greasy developing ink and then washing away the unhardened portions of the coating which carries the ink with it leaving the hardened details still carrying the ink.

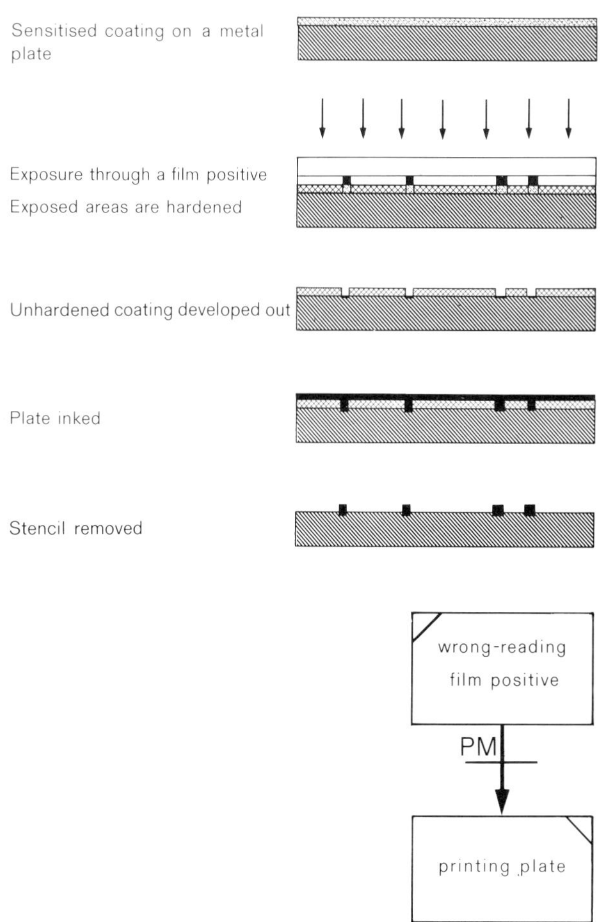

Fig. 5.34 Steps in making a deep-etch printing plate

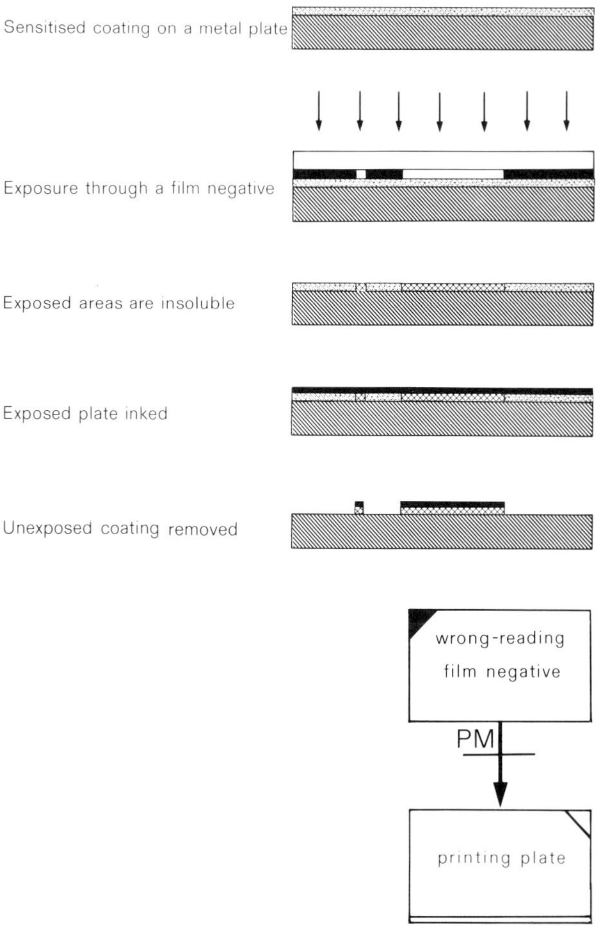

Fig. 5.33 Steps in making a surface printing plate

Deep-etch plates (Fig. 5.34) are made from positives which can be either photographic or drawings done on transparent plastic, providing the drawing is suitably opaque to withstand light. When exposed, the light-hardened coating serves as a stencil that protects the non-printing areas, while the image areas are chemically etched into the surface of the plate slightly below the non-printing areas. The plate is then developed in the normal manner by rubbing it over with ink which, in effect, puts the ink on the plate through the stencil, after which the hardened stencil is removed.

Bimetal plates (Fig. 5.35) consist of two different metals, one for the image areas and one for the non-printing areas. Practically all bimetal plates today use copper which is receptive to ink for the image areas and aluminum, chromium or stainless steel which is water receptive for the non-printing areas. For negative working plates, the image metal is electroplated onto a sheet of non-image metal. After normal processing, the copper is etched away from the non-print areas exposing the water-receptive base metal. For positive working plates the metals are reversed.

5.10.1 Copying of press stencils for silk screen printing

The silkscreen process is a form of stencil printing. The stencil is formed on a silk screen fabric that has been tightly stretched over a frame. The areas that are to remain unprinted are blocked out with

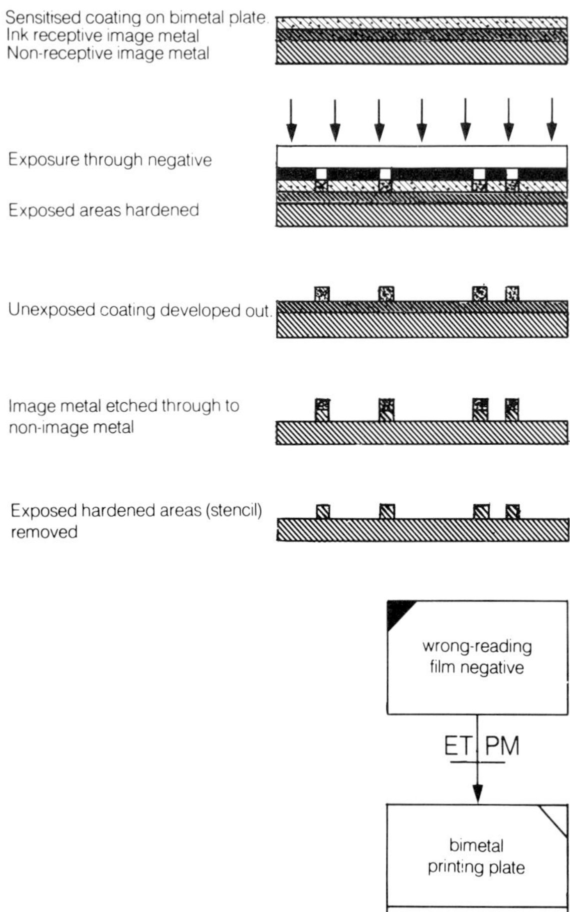

Fig. 5.35 Steps in making a bimetal printing plate using a negative

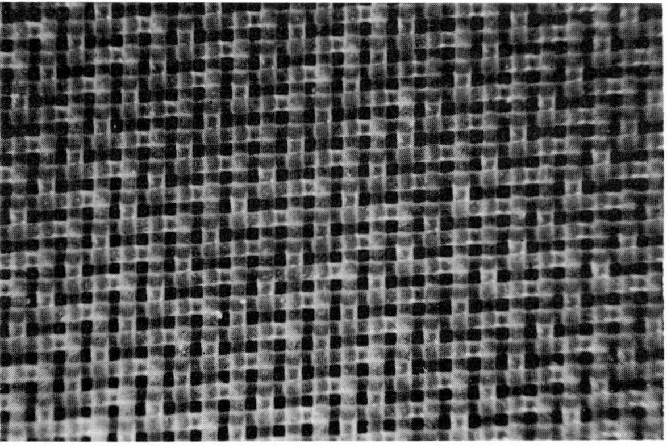

Fig. 5.36 Enlarged monofilament polyester fabric with 150 threads per cm (210× magnification)

a masking material that is impervious to an oil base printing ink. The image is printed by squeezing the ink through the open areas of the silk onto almost any kind of surface.

For cartographic purposes, the screens that are normally used are woven with polyamide (nylon) or polyester threads, which, because of their dimensional stability, tensile strength and resistance to abrasion and chemicals, make them ideal for precise registration in multicolour printing (Fig. 5.36). The smooth monofilament threads allow the ink to penetrate the screen easily so as not to block the mask in the printing area of the stencil. Synthetic silk screen with 130 to 160 threads per cm are usually suitable for printing fine map details. Fabrics that have been dyed yellow or red will prevent the scattering of actinic light when exposing an original image directly onto the light-sensitive surface of the screen. The image used to produce a stencil must be in positive form on a transparent plastic base or photographic film.

There are three photomechanical methods used to produce silk screen stencils: the direct method, the indirect method and the combined method.

In the direct method (Fig. 5.37), a right reading positive original is exposed directly onto the surface of the screen. The screen itself is coated with a dichromated colloid, such as gelatin, polyvinyl alcohol, polyvinyl acetate (see 5.4) or with polymers sensitized with a dichromate, a diazonium compound or an azide. The coating is normally applied evenly, by hand, to both sides of the screen using a round-edged spreader. The coating is then dried. After being exposed, the image is developed by washing away the light-sensitive coating from the unexposed areas of the stencil with warm water leaving only the exposed light hardened stencil.

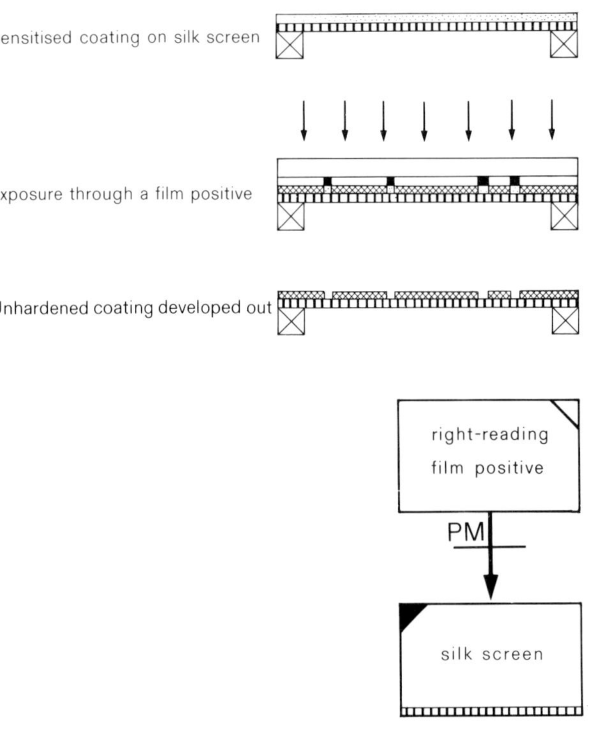

Fig. 5.37 Direct method of preparing a silk screen stencil

In the indirect, or transfer method (Fig. 5.38), the stencil is prepared independently from the screen by exposing a wrong reading positive image onto a special presensitized photostencil film which is later transferred to the silk screen. After exposure, the unexposed areas of the film are washed away with hot water until they are free of the emulsion and completely clear. The remaining light hardened stencil is then transferred on to the screen by placing the film, emulsion side up, on a supporting backboard and lowering the screen onto the film. Since the film is already wet, it will adhere to the screen and, when dry, the supporting film base can be stripped away leaving the stencil attached to the screen.

In the combined method, a non-sensitive synthetic film layer, on a polyester supporting base, is coated with a light-sensitive polymer solution. The coating is applied through the silk screen mesh onto the film with the aid of a squeegee. When dry, the supporting polyester base is stripped away, leaving the coated light-sensitive film layer attached to the screen. A right reading positive original is then place in direct contact with the sensitized screen and exposed in a vacuum frame in the normal manner. Following exposure, the image areas are removed with water, leaving the light-hardened synthetic film stencil that blocks out the non-image areas.

In all three cases, the ink is squeezed through the open image areas of the stencil onto the appropriate material.

Because synthetic silk screen fabric is extremely strong, the same screen can be used over and over again as the stencil can be removed with a solvent.

Some applications of the silk screen process in

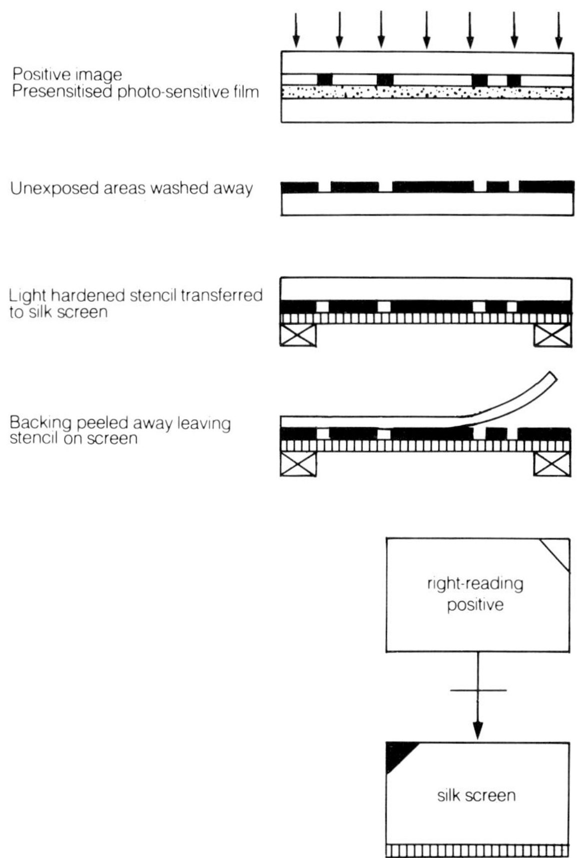

Fig. 5.38 Indirect or transfer method of preparing a silk screen stencil

cartography, include the production of multicolour maps in small quantities, overprinting information on existing maps, production of graphic materials for application by the dry transfer technique (see 1.5), for printing with fluorescent inks and to print maps on cotton, glass, wood etc.

CHAPTER SIX

Geometrical Image Transformations

6.1 Line photography (including cameras and microfilm reproduction)

Line photography is used for black and white and multicolour line originals that do not require the use of half-tone screening for tonal reproductions.

The quality of the original image is of the utmost importance. All line and solid details must appear dense and sharp on a clean white or transparent background. Colour also is an important factor. Because the type of film used in line photography will not record all colours as black or white, it is necessary to use colour filters when photographing certain single and multicolour line originals.

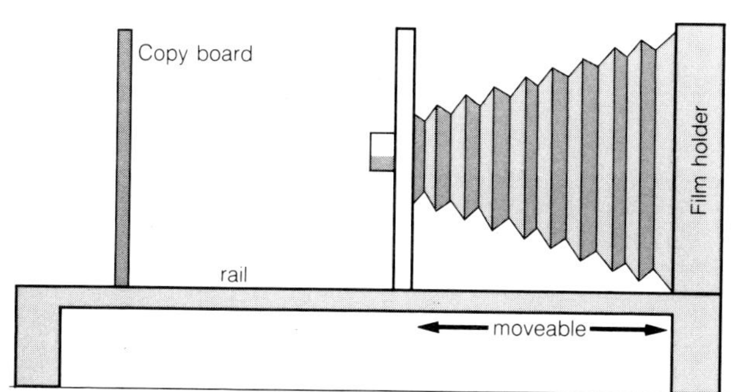

Fig. 6.1 Gallery or one room camera

The type and size of camera used in cartographic operations will vary according to the requirements of each organization. They may range from the fairly small type cameras with a maximum film size of approximately 40 cm × 50 cm (16 in × 20 in) to extremely large cameras capable of producing negatives up to approximately 122 cm × 183 cm (48 in × 72 in). There are generally two types of cameras used in cartography: the gallery or one room camera (Fig. 6.1) and the darkroom or two room camera (Fig. 6.2). The gallery cameras are independent from the darkroom facilities and the light proof film holder must be transported between the camera and darkroom for the loading,

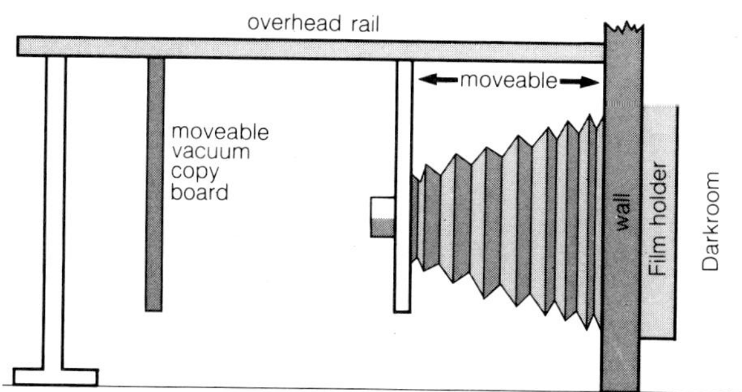

Fig. 6.2 Darkroom or two room camera

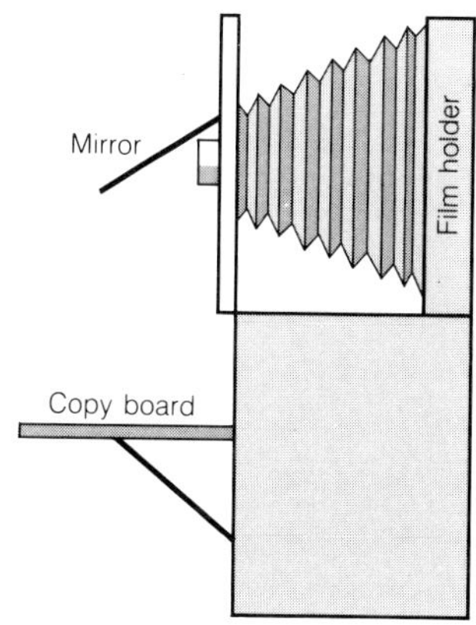

Fig. 6.3 Vertical type camera

exposure and developing operations. Darkroom type cameras are more sophisticated and, as the name implies, they extend into a darkroom from where, except for the loading and removal of the film, all operations are carried out by the operator through a central control panel. There are also vertical type cameras (Fig. 6.3) the bodies of which are usually horizontal with a parallel copy board. Since the focal plane of the camera and the copy plane are at right angles to each other, it is necessary to photograph the original image through a prism or mirror. This automatically produces a right-reading negative from a right reading original image. Such reversals can also be made with the

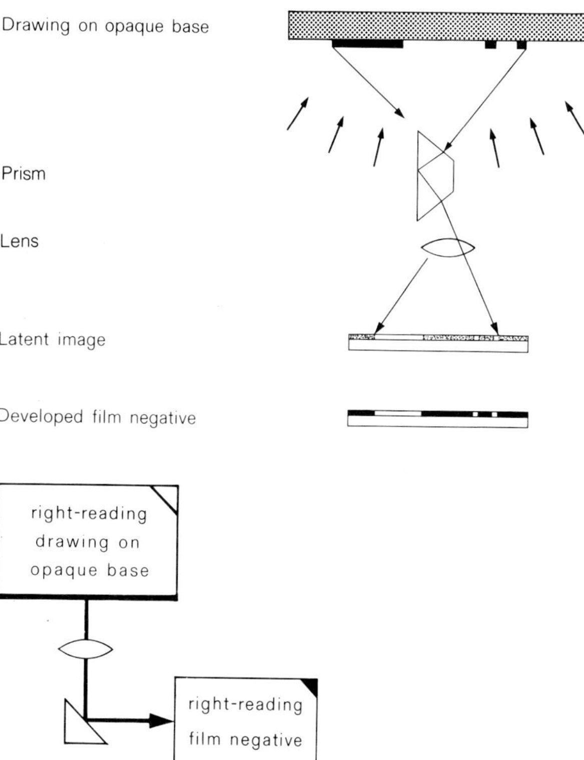

Fig. 6.4 Exposure arrangement through prism to produce a right-reading negative

gallery and darkroom type cameras by photographing through a prism (Fig. 6.4).

Before photography, the exact dimensions of the original and the scale, or size, of the required reproduction negative must be calculated to establish any necessary enlargement or reduction factors. The original is then attached to the copy board and the camera and board are focused at the required distance apart to obtain the correct size of negative. The lighting angle and the lens aperture are adjusted to give the best definition and resolution of detail and the original is exposed for a calculated time or measured amount of light. Following the exposure, the latent image on the film is developed in the normal manner (Fig. 6.5).

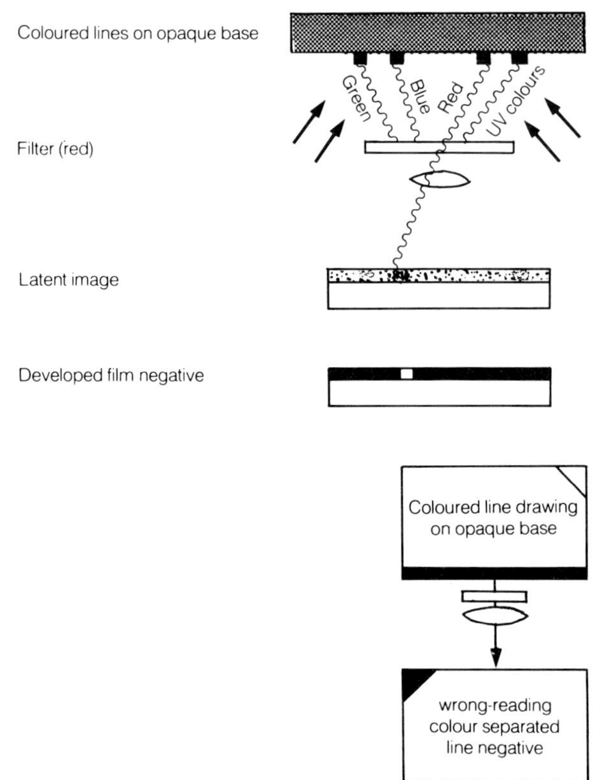

Fig. 6.6 Exposure arrangement for colour separation of line detail

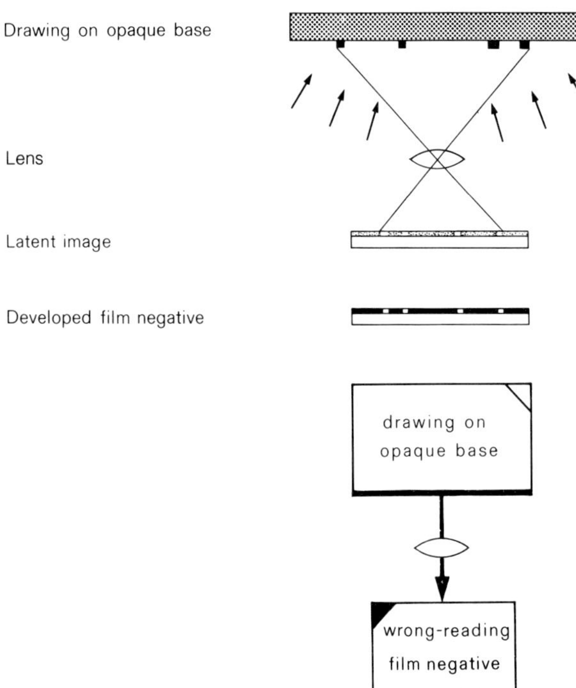

Fig. 6.5 Exposure arrangement to produce a negative image from a positive original

If the different colours in a multicolour line original are to be photographically separated, the original is photographed through an appropriate colour filter placed in front of the camera lens. The purpose is to obtain a clear image on the negative of the colour that is to be retained and to have all other colours covered by the black opaque background of the developed negative. The filters themselves will transmit light of their own colour while absorbing light of other colours. A red filter, for example will transmit red light but absorbs green, blue and ultraviolet colours (Fig. 6.6). So by using the appropriate filter it is possible to filter out, or reduce, unwanted colours. In practice, however, it is not always possible totally to hold back an unwanted colour; consequently, some retouching with opaque is usually required on the negative.

Multicolour line originals can also be colour separated by electronic scanning (see 7.1.2).

High resolution camera lenses, combined with high resolution photographic film, make it possible to reduce considerably the scale of a map original onto 35 mm or 70 mm film and subsequently to enlarge this microfilm back to its original scale without much loss of quality or original scale. This is of particular value where large numbers of records must be stored, or where a particular map must be retrieved quickly and reproduced at its original scale on demand. Large scale monochrome maps or plans are particularly suited for this purpose.

6.2 Continuous-tone photography

The procedure for making continuous-tone negatives or positives, is basically the same as for making line negatives as described in section 6.1, except, in this case, the objective is to obtain the best possible density range between the deepest black and the brightest white of the original document. The type of film normally used is a medium contrast film, referred to as continuous-tone film, which records all the gradations of tone contained in the original.

A continuous-tone negative of an opaque or transparent continuous-tone original can be utilized as an original to produce prints on various kinds of photographic papers, or to produce screened contact positives (see 7.1.3) for half-tone printing.

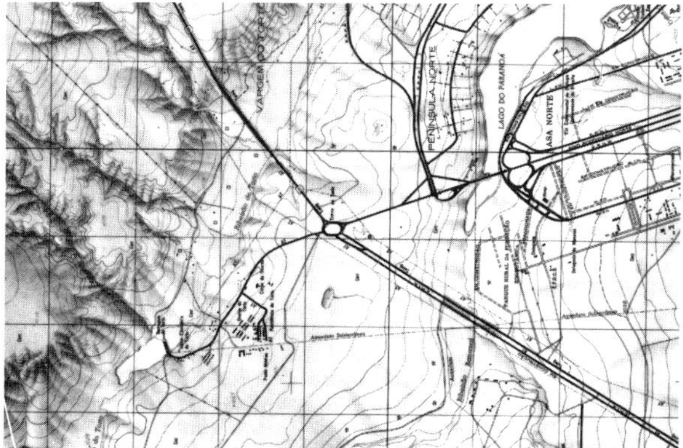

Fig. 6.8 Reduced black and white positive of a multicolour map for use as a guide copy (reproduced in half-tone)

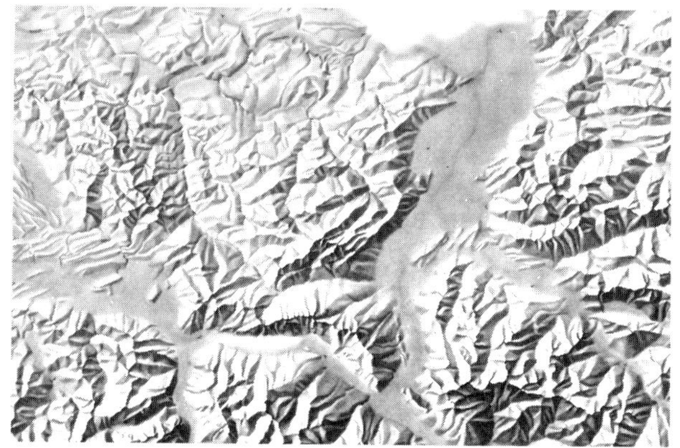

Fig. 6.7 Reduced shaded relief drawing (reproduced in half-tone)

The main applications of the process in cartography are in the making of rectified and unrectified aerial photographic prints for construction of photomaps and to enlarge, or reduce, shaded relief drawings (Fig. 6.7). The assembled photomaps or shaded relief drawings then become the originals which are photographed in half-tone for printing in their final form. Often black and white continuous-tone photographs of multicolour maps are used as guide copies for the compilation of other maps (Fig. 6.8).

The quality of the original continuous-tone photograph will greatly influence the quality of the final printed image, hence, sometimes, originals have to be treated to improve their reproduction qualities. It may be necessary to re-photograph the

original to achieve the desired effect, or to retouch the original before photography to either emphasize or reduce the tone or detail in certain areas. The most effective means of accomplishing this is with an airbrush, which in skilled hands is capable of producing all the tonal gradations of the original document.

6.3   Colour photography

The process of making colour photographs is much the same as for making black and white continuous-tone photographs (see 6.2), except that special colour processing equipment and materials are required and the type of film used is equally sensitive to all colours of light.

Colour photography is not normally a part of the map reproduction process. However, it is sometimes used to reduce or enlarge coloured shaded relief drawings and colour air photographs used in the production of photomaps that are subsequently colour separated for printing in their final form. There is also often a requirement to prepare positive transparencies of multicolour maps for projection purposes.

CHAPTER SEVEN

Value Transformations

7.1 Half-tone reproduction

The half-tone reproduction process is used to reproduce black and white or coloured original copies. Since most printing presses can only print one strength and density of colour, it is necessary to convert the tonal range of the original into a series of dots. There are two basic methods of achieving this: one by the use of a glass cross line screen in the back of the camera and, two, by the use of magenta or grey contact half-tone screens. Both methods are described in 7.1.1 and 7.1.3.

Although half-tone dots are not visible at the normal viewing distance, each dot is a separate piece of line work which will reproduce a single speck of colour varying in size from larger dots in the shadow areas of the copy to tiny dots in the light areas. The visual difference in tones is created by the amount of printing ink which each dot carries in proportion to the amount of white paper that surrounds it, therefore creating the illusion of a continuous-tone copy (Fig. 7.1).

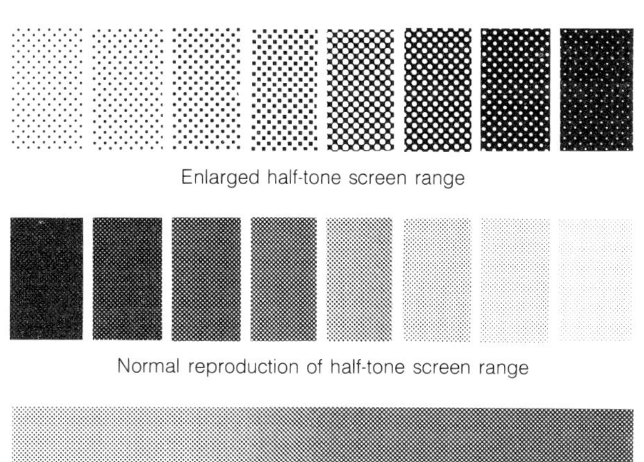
Enlarged half-tone screen range

Normal reproduction of half-tone screen range

Fig. 7.1 Half-tone screen range

Continuous-tone originals can be in many forms. However, in cartography the most typical are black and white or coloured air photographs and satellite imagery used to produce photomaps, and shaded relief drawings which have been either rendered in monochrome or in colour. Coloured originals contain many areas of distinctly different colours which it would be impossible to print separately. However, most colours can be reproduced by some mixture of the three process colours, i.e. yellow, magenta, and cyan with the addition of some black.

To reproduce a coloured original, it is necessary to separate the colours of the copy into four separate photographic half-tone images (yellow, magenta, cyan, and black). This is accomplished by exposing the original through colour filters so that each separate negative contains the proportional amount of colour to be printed. In order to prevent the possibility of a pattern or moiré effect occurring during printing, the half-tone screen is rotated to a different angle for each separation so as to place the dots of the four colours at angles whereby they will not create unwanted patterns. This condition is usually satisfied when the screen angles of the black, magenta and cyan are 30° from each other, with the yellow at an angle of 15° between two darker colours (Fig. 7.2).

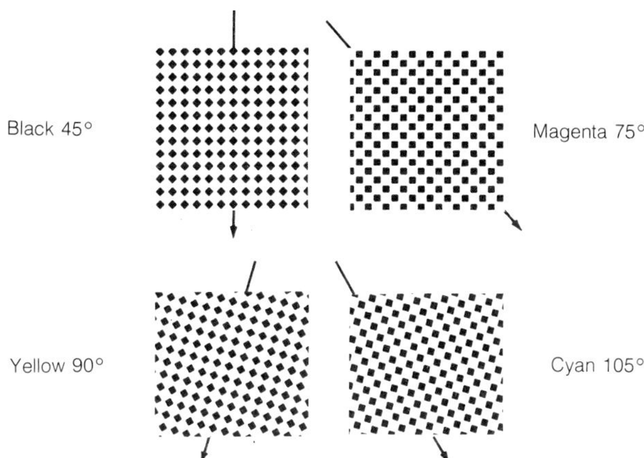

Fig. 7.2 Half-tone screen angles for four-colour printing

7.1.1 Half-tone reproduction by photography

A half-tone is produced by exposing a monochrome or coloured continuous-tone copy through a half-tone screen mounted in the back of a camera. There are two basic types of half-tone screens: the glass circular type mounted in the back of the camera (Fig. 7.3) and the magenta or grey contact screens which are normally used in a vacuum frame but can also be used in the back of the camera providing the camera is equipped with a vacuum back (Fig. 7.4). In general, the magenta and grey contact screens (see 7.1.3) are the most popular as they have the advantage of being reasonably inexpensive and are both faster and easier to use.

A glass screen consists of fine lines that have been cross ruled at right angles to each other to form tiny squares (Fig. 7.5). When exposed, these squares break up the light and dark tones of the original into a series of dots which normally range in value from 70% to 95% in the darkest tones, 35% to 70% in the mid-tones, and to as low as 5% in the lightest areas (Fig. 7.6).

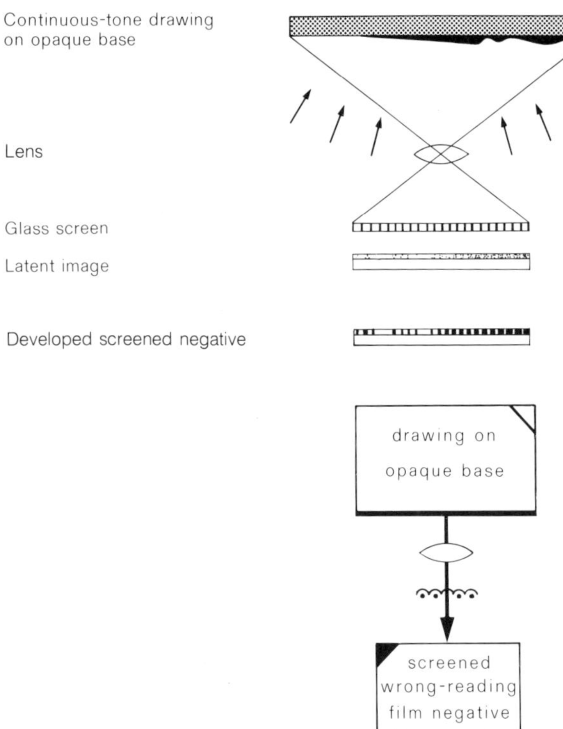

Fig. 7.3 Photographic sequence using circular glass screen mounted in back of camera

Screens are classified according to the number of lines to the linear cm or inch. The most commonly used are the 48 or 54 lines per cm, or the 120 or 133 lines per one inch screens, although screens can be obtained up to 120 lines per cm or

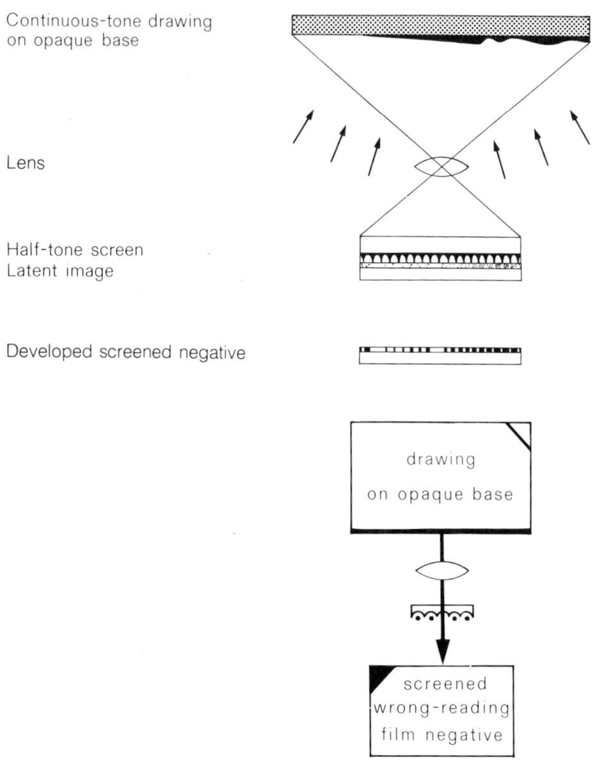

Fig. 7.4 Photographic sequence using a magenta or grey contact screen

Fig. 7.5 Enlarged cross-line screen

300 lines per inch. When a glass cross-ruled screen is used, the proper dot formation on the exposed film is dependent upon the correct adjustment of the screen distance (from the film in the back of the camera) in relation to the lens aperture.

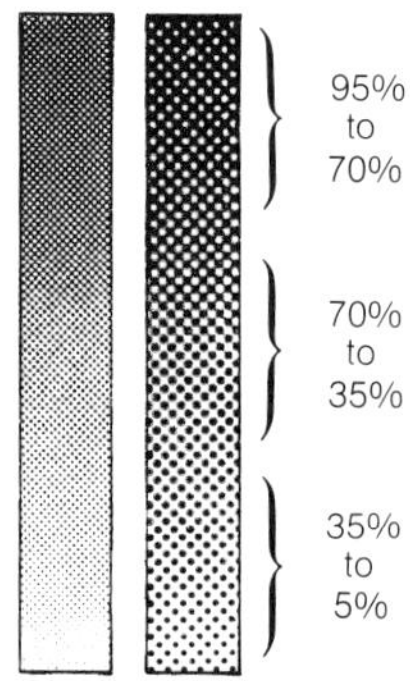

Fig. 7.6 Tone range of screen

On exposure, the light spread around the screen rulings can be either increased or decreased according to the aperture opening of the lens. A large aperture opening will cause the light to spread resulting in a small highlight dot on the negative, whereas reducing the size of the aperture will reduce the spread of light and the dot size proportionally. Because of the relationship of aperture opening and dot size, the use of multiple stop exposures (highlight, middle tone and detail) for glass screen work is generally accepted for photographing shaded relief drawings and continuous-tone photomaps. The highlight stop, which is the widest lens opening, reduces the size of the highlight dot and, if desired, can be used to drop-out the dot completely from the highlight areas. The middle tone stop (approximately one full stop less than the highlight stop) records the intermediate tones of the copy and the detail stop (approximately one full stop less than the middle tone stop) is used to record the fine details in the darkest tones of the copy. A supplementary 'flash' exposure to light reflected by a white sheet of paper or flash lamp is used to increase the opacity of

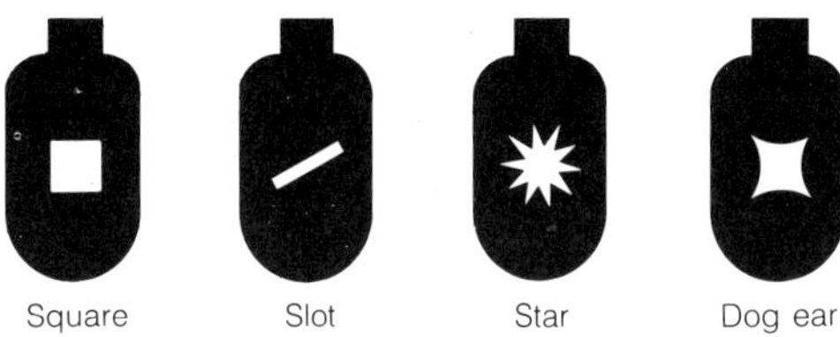

Fig. 7.7 Some typical diaphragms

each dot without increasing its size or shape. This exposure is made with a small lens aperture approximately one stop smaller than was used for the shadow tones.

Different diaphragm shapes (Fig. 7.7) are often used in conjunction with the camera lens to vary the dot formation; for example, a square shaped diaphragm will produce square dots throughout the entire tonal range. An elongated slot will produce an elongated chain of dots.

Glass cross-line screens are generally used to make direct separation negatives from coloured copies as the screen can be conveniently rotated in its holder in the back of the camera to provide the necessary screen angles. The copy is separated into four separate negative images, i.e. yellow, magenta (process red), cyan (process blue) and black. To achieve this, the negatives are exposed through a blue, green or red filter. For instance, if the copy is photographed through a blue filter the resultant negative becomes the yellow printer. Similarly, the green filter will produce the magenta negative printer, the red filter will produce the cyan negative printer, and the three filters combined will produce the black negative printer (Fig 7.8). For each exposure the glass screen is rotated to the appropriate screen angle, i.e. yellow 90°, magenta (red) 75°, cyan (blue) 105°, and black 45°.

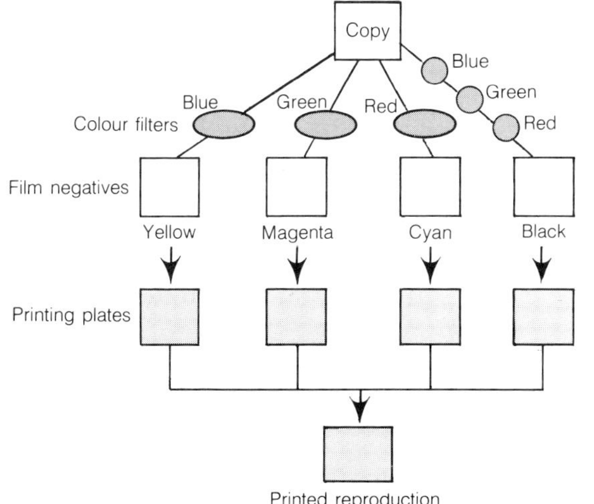

Fig. 7.8 Sequence of direct four colour separation

7.1.2 Half-tone reproduction by scanning

Automatic colour separation and colour correction is accomplished with the use of electronic scanning devices (Fig. 7.9) which are capable of producing sets of colour separated half-tone positives or negatives in one operation.

Fig. 7.9 Electronic scanner

The basic steps involved are as follows. The coloured image is attached to a rotating cylinder. As this cylinder rotates, a light beam automatically moves column by column and point by point across the face of the image. If the original is a colour transparency, the light is transmitted through it, or if the original is opaque, the light is reflected from it. In either case, the light passes through a beam splitter and colour filters and is received by the photo-cells which generate an electric signal, the strength of which depends on the intensity of the light received. The signal is then processed through the electronic feed-back controls to the modulated light projectors and exposed point by point onto film to produce the colour separated half-tone negatives or positives (Fig. 7.10).

As practically all the colour components of a map are produced separately, colour separation by electronic scanning is seldom used in cartography. It can, however, be used to colour separate shaded relief drawings that have been produced in colour, coloured aerial photographs and printed multi-colour maps for half-tone printing.

7.1.3 Half-tone reproduction by contact printing

Contact screens are used to convert the highlights, middle tones and shadow areas of a continuous-tone negative to a series of dots for half-tone reproduction. The contacting is usually done in a vacuum printing frame in which the screen is

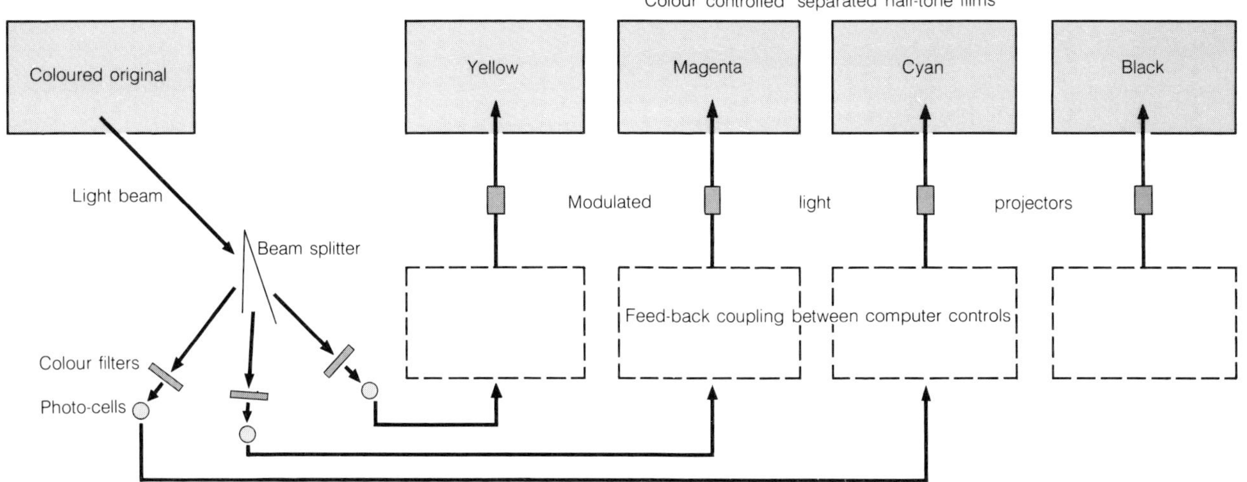

Fig. 7.10 Schema of a scanner

exposed with its emulsion side in direct contact with the emulsion side of a lith type film. Contact screens can also be used in a camera providing the camera is equipped with a vacuum film holder.

The screens are photographically manufactured on film and consist of a regular pattern of dots. Each dot in the pattern is vignetted (Fig. 7.11), having a maximum density in the centre graduating to a minimum density around its edge. The difference between the maximum and minimum densities determines the density range of the screen (Fig. 7.12). The centres of the dots are equidistant and the size of the screen is measured by the number of dots per cm or inch. Some typical sizes are 48, 54, 60, 70 and 80 dots per cm equal to 120, 133, 150, 175 and 200 dots per inch.

During the contacting process, the dot formation on the resulting contact negative depends on the amount of light that penetrates the vignetted area of the dot. A large amount of light will penetrate almost all of the vignetted area except for the small solid point in the centre which will remain as a

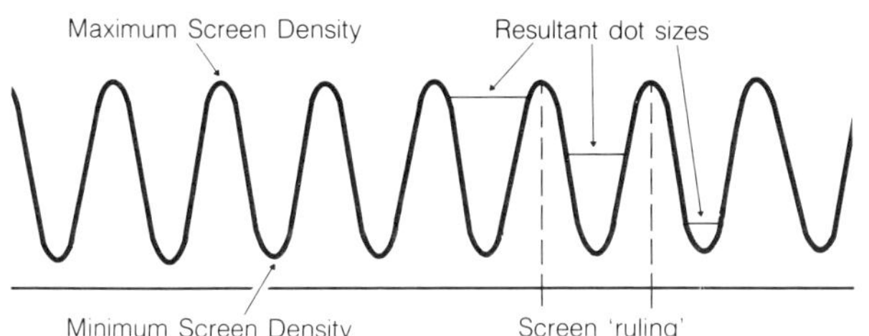

Fig. 7.12 Cross-section of vignetted screen densities (After J.S. Keates)

small clear point on the contact negative. A small amount of light will only penetrate the less dense periphery of the dot leaving a small dense point in the negative. The printed image will be the reverse of this so that the highlight areas will contain the small solid dots graduating to large dots in the shadow areas.

When tints of around 50% have to be differentiated, preference is usually given to using the elliptical or chain like form of dots (Fig. 7.13)

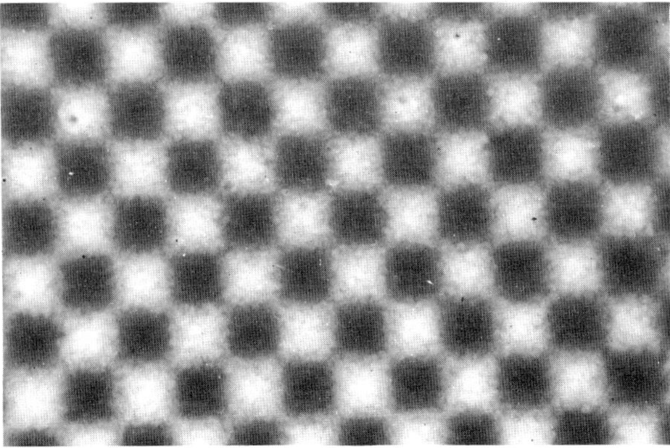

Fig. 7.11 Structure of vignetted magenta contact screen (enlarged)

Fig. 7.13 Half-tone image formed by elliptical or chain dot screen (enlarged)

which eliminates the abrupt gap which is visually noticeable with the square dots.

There are, basically, two types of half-tone screens. The grey screen which is mainly used for specialized direct colour separation and the magenta screen which is extensively used both for black and white half-tone reproductions and for making contact half-tone positives from continuous-tone colour separated negatives.

The magenta contact screens are the most commonly used, as the magenta colour lends itself to certain modifications through the use of colour filters during exposure. A yellow filter will lower the contrast on the contacted copy, whereas, a magenta filter will greatly increase the contrast (Fig. 7.14). A partial exposure through either filter will give intermediate results. This allows for the tonal characteristics of the contact screen to be adjusted to the density range of the negative or positive to be screened.

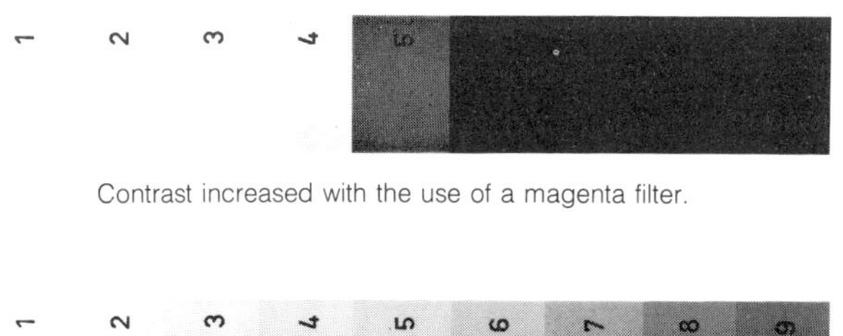

Contrast increased with the use of a magenta filter.

Contrast lowered with the use of a yellow filter.

Fig. 7.14 Tonal modifications using colour filters with a magenta screen

In cartography, contact screens are mainly used for the reproduction of orthophoto maps (Fig. 7.15) and similar continuous-tone documents which are to be printed by either offset, silkscreen, xerography or diazo techniques.

Fig. 7.15 Portion of an orthophoto map produced with a 54(133) magenta dot screen

7.1.4 Screenless lithographic printing

Unlike the half-tone reproduction techniques described in the preceding sections, screenless printing is a method of reproducing, by conventional lithographic printing, a continuous-tone image without using a half-tone screen.

To accomplish this, it is necessary to break down the continuous-tone image into minute ink receptive spots, on the printing plate. These spots are randomly distributed throughout the printing image and vary in size and shape according to the grain of the plate and the tonal range of the continuous-tone negative or positive.

The ability of the printing plate to produce suitable ink receptive spots is dependent on the size of the peaks and valleys on the grained printing plate (Fig. 7.16) and on the thickness of the light-sensitive coating on the plate. For this purpose, the commercially available presensitized diazo positive and negative working aluminium plates have been found to be the most suitable, as they have a very fine and homogeneous grain structure and a very thin light-sensitive diazo layer. These combined properties provide a printing resolution in excess of 200 lines per cm or 500 lines per inch.

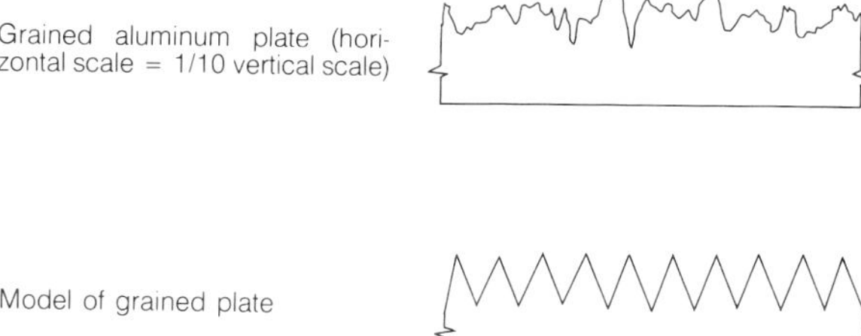

Fig. 7.16 Enlarged cross sections depicting random peak and valley structure on grained plate and simplified model as a series of cones (After I. Pobboravsky and M. Pearson)

A plate can be made from either a positive or negative continuous-tone original (Fig. 7.17). In either case, the image is placed in direct contact with the light-sensitive layer of the aluminium plate in a vacuum printing frame and exposed to a light source. The amount of light that passes through the original to form the latent image on the plate is controlled by the tonal densities of the negative or positive that is used. If a negative is used, following the development of the latent image, the light hardened negative working coating remains anchored, at various levels, to the peaks of the grain on the plate to become the ink receptive

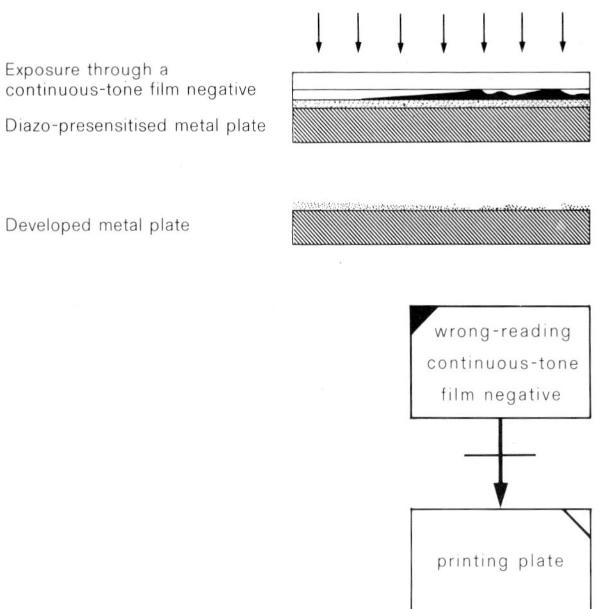

Fig. 7.17 Exposure sequence using a continuous-tone negative

printing image (Fig. 7.18). The darker tones are represented by the larger spots and the lighter tones by the smaller spots. Similarly, if a positive is used, the insoluble coating, following development, remains anchored in the valleys in the grain to form the ink receptive image (Fig. 7.19).

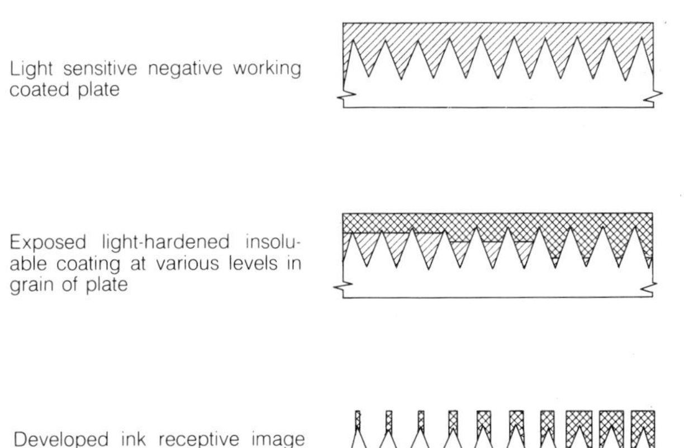

Fig. 7.18 Negative working presensitized plate (After I. Pobboravsky and M. Pearson)

Screenless printing can be carried out on a conventional lithographic offset printing press to produce either single- or multicoloured copies using conventional inks and fountain solutions.

Screenless printing, because of its ability to reproduce very fine details, is used to produce orthophoto and photomaps in which the retention of the original photographic detail is of the utmost

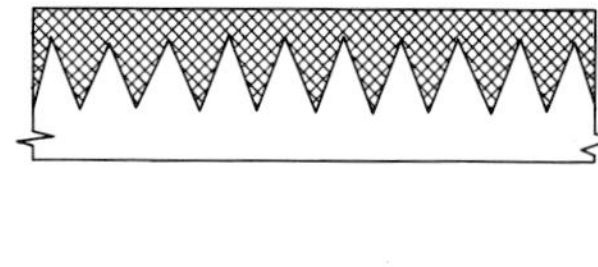
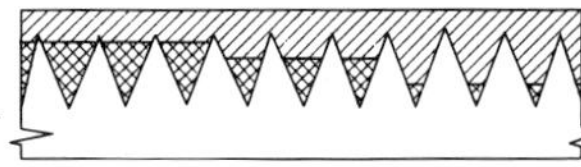
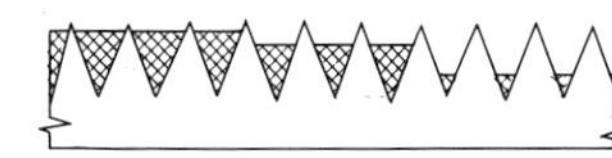

Fig. 7.19 Positive working presensitized plate (After I. Pobboravsky and M. Pearson)

Fig. 7.20 Example of screenless printing

importance. Figure 7.20 is an actual example of screenless printing.

7.2 Hard dot contact printing (tint screening)

In the lithographic printing process it is not possible to print different tones of the same colour as a continuous-tone image. Such tones can only be simulated with the use of so-called hard dot or line screens that make it possible to reproduce a variety of tonal ranges for the printing of uniform tinted areas.

The characteristics of these screens on which the hard dot or line elements are distributed in regular or quasi-regular patterns are:

a) form of the screen element (line, dot, symbol, etc.)

b) coarseness of the screen or spacing between the screen elements

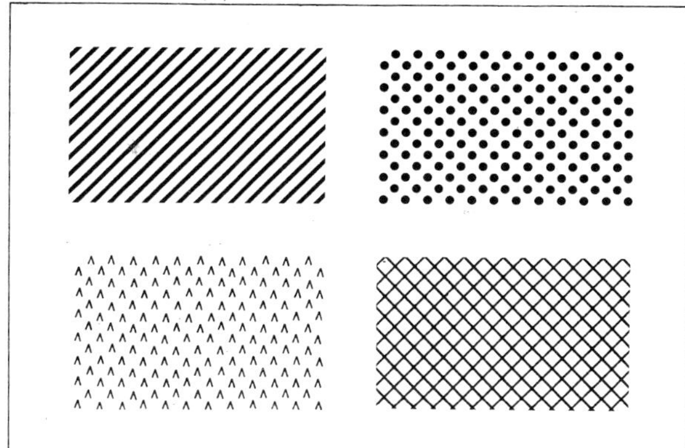

Fig. 7.21 Examples of dot, rulings and patterns

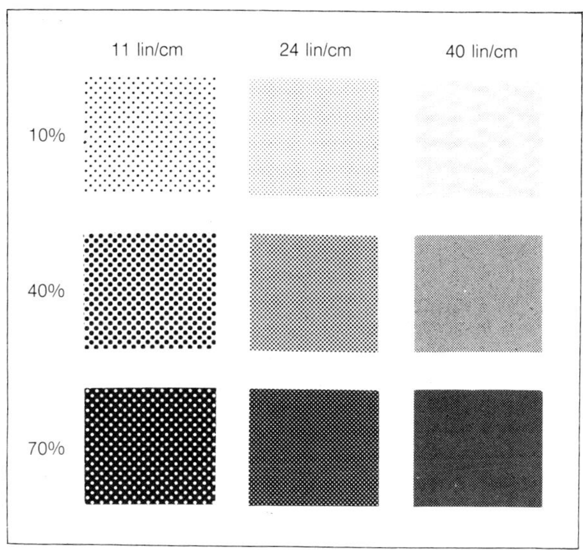

Fig. 7.23 Dot screen percentage values

c) screen values in percentage of coloured area per area unit
d) screen angles
e) symmetry of the screen or pattern

There are many different types of screen available with dot, line and cross-rulings or, with various regular or randomly distributed symbol patterns (Fig. 7.21). The dot screens are available in checker-board, square, and round patterns (Fig. 7.22).

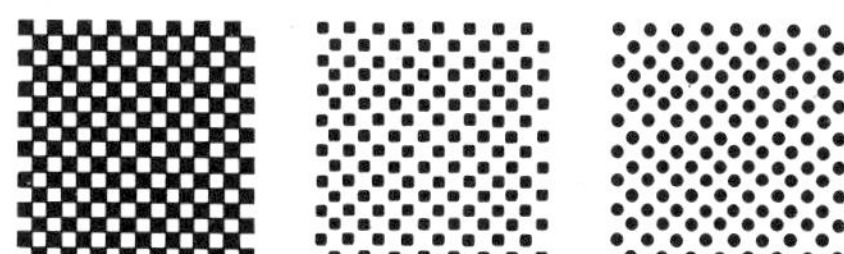

Fig. 7.22 Checker-board, square and round dot patterns

The coarseness of a screen is normally expressed by the number of screen elements per cm or inch. Dot screens are available up to 120 lines per cm or 300 lines per inch, but printing methods and paper quality limit their application. For offset printing, the finest screens normally used have 60 lines per cm (150 lines per inch). Screens that are coarser than 40 lines per cm (100 lines per inch) clearly show their structure.

The screen value is expressed in percentages of the opaqued area of an area unit (Fig. 7.23). Commercially available line or dot screens normally have the following percentages: 5, 10, 20, 30, 40, 50, 60, 70, 80 and 90%. For equal-interval value scales, however, additional screens, such as 6, 16, 25, 34, 44, 55%, are usually recommended, depending on the individual printing colour.

As explained in 7.1., the angle at which a screen is printed is very important as it can cause various visual effects. For instance, a screen with a promi-

nent vertical pattern may appear to be less even than those with other orientations. Special care must also be taken when two or more dot screens are overprinted. If the screens cross each other at angles less than 15° (approx.), a visually disturbing pattern, known as a moiré pattern, will appear (Fig. 7.24). The smaller the angle of intersection between two dominant orientations, the more severe the moiré effect will be. The optimum screen angles for multicolour screen combinations are illustrated in Fig. 7.2. The angle of a screen is determined by the angle between its main orientation and the vertical direction measured in a clockwise direction. The darkest colour is normally screened at an angle of 45°. All additional dark colours should have an orientation that is different by 30°. Lighter colours, such as yellow, may be put at angles between.

The standard application of hard dot contact printing is in the screening of areas that are to

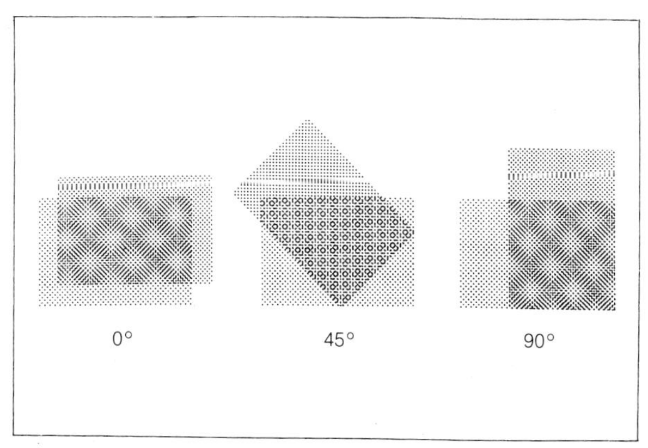

Fig. 7.24 Moiré patterns

contain a certain colour tint on the printed map. A separate mask is required for use along with the appropriate master screen for each different tint. Special care must be taken when preparing masks, because scratches, dirt or dust will result in imperfections in the printed image which are especially noticeable in a regular pattern.

The contact printing of screens (Fig. 7.25) is done in a vacuum printing frame in which the emulsion side of the screen is placed in direct contact with the light-sensitive side of the material on which the image is to be transferred. The mask is then registered over the screen and the assembled material is exposed and developed in the normal manner.

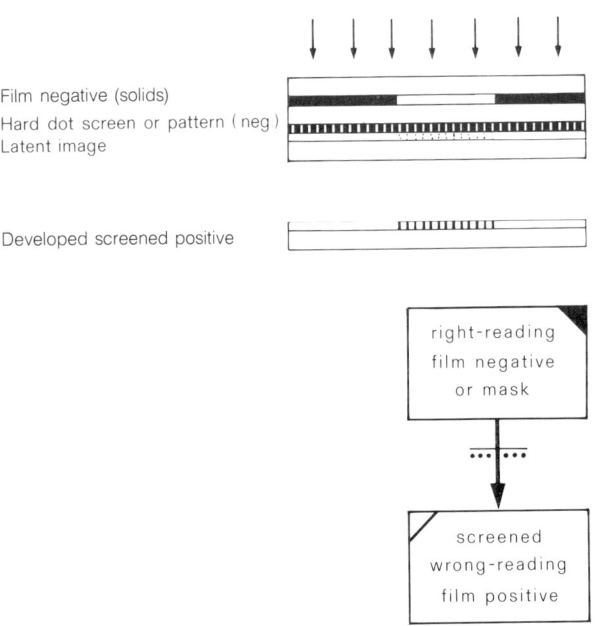

Fig. 7.25 Exposure sequence using a negative mask

Combined images of several different screens, provided they are in negative form, can be obtained by exposing them one after the other along with their respective negative masks, onto the same sheet of film, plastic or printing plate, using a punch register system (see 3.2).

If only one area has to be screened, another procedure, using a direct positive film (5.1.2.) along with a positive mask, can be applied as illustrated in Fig. 7.26. The first exposure with the master positive screen or pattern provides a latent image over the whole film. In the second exposure, the screen is replaced by the mask. During exposure, the light destroys the latent screen image outside of the masked area. The latent image protected by the mask, is then developed.

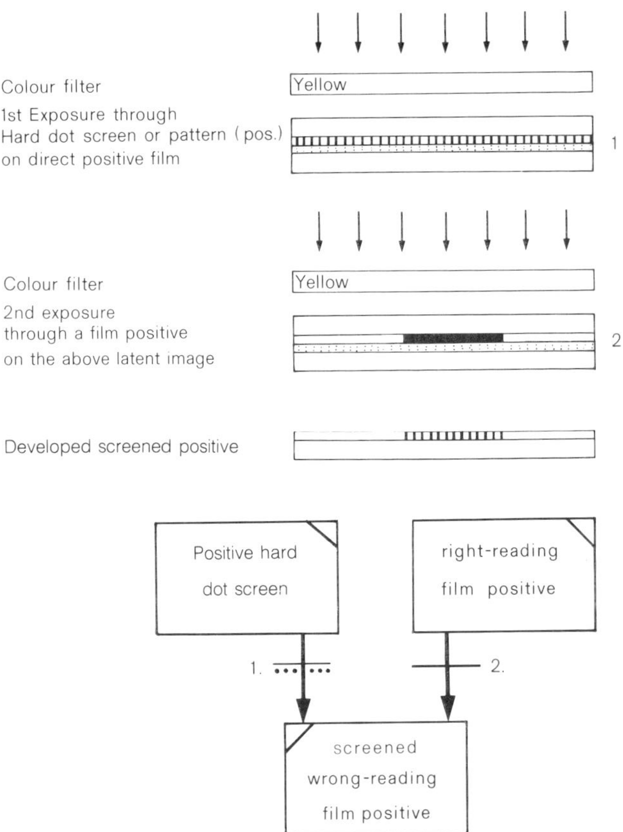

Fig. 7.26 Exposure sequence using direct positive film

7.3 Changing line widths

Normally, copies are produced which exactly duplicate the same line widths or dot sizes as those of the original image. In some applications, changing the line width or the dot size may be useful, for example in:

—retaining the original widths of lines in spite of photographic enlargement or reduction
—eliminating the thinnest lines of a drawing while decreasing the widths of other line details
—reducing or enhancing the tonal values in a screened image
—reducing or thickening lines for better graphic contrast in relation to other map elements
—preparing outlines (see 7.4)
—edge enhancing (see 7.10)
—the pictoline technique (see 7.11)
—the block-out technique (see 7.12)

There are three different techniques that can be used for changing the widths of line or dot detail, which are briefly outlined as follows.

In a process camera, a device in the form of an inclined rotating glass disc is placed in front of the lens. The light rays describe small rotations on

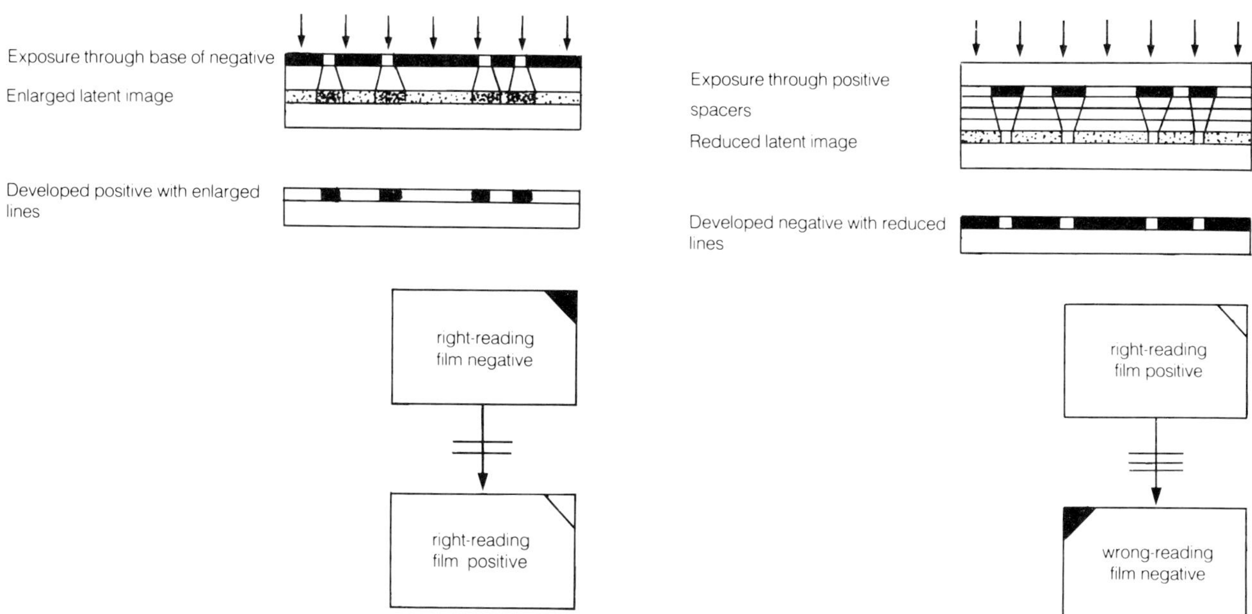

Fig. 7.27 Contact copying through base of film negative to enlarge line width

Fig. 7.29 Contact copying through spacers to reduce line weight

the photographic film during the exposure. The amount of tilt of the glass disc influences the amount of increase or decrease in the line width.

In contact printing, the distance between the original and the copy material influences the line width. Therefore, a controlled increase or decrease in line width can be obtained by copying through the base of the original (Fig. 7.27) or, by placing transparent diffusing sheets or a spacer between the original and the copy. For heavier lines, exposure is made through a negative (Fig. 7.28)

for thinner lines the exposure is made through a positive (Fig. 7.29).

Some photographic films also allow for an increase or decrease in line width by varying the exposure time.

7.4 Preparation of outlines

Outlines of areas can be prepared from opaque solid originals, such as area masks and similar solid map details (Figs 7.30 and 7.31) by the photographic contact printing process using a direct positive film (see 5.1.2) and an excessive exposure as described in subject 7.3. The excessive exposure causes the light to spread under the edges of the solid areas, which, following the exposure and developing operations, results in an outline only

Fig. 7.28 Contact copying through spacers to enlarge line width

Fig. 7.30 Positive original (solids)

Fig. 7.31 Outline copy

Fig. 7.33 Combined line and screen image

of the area. If a positive original is used, the lines will be located slightly inside of the area and if a negative is used, the lines will appear slightly outside of the area.

Figure 7.32 illustrates the copying arrangement for preparing outlines using a positive original.

The original solid image is assembled in a vacuum printing frame in direct contact with the emulsion side of the direct positive film. The assembled material is then given a longer than normal exposure through a yellow filter which destroys the film's built-in maximum density in the exposed

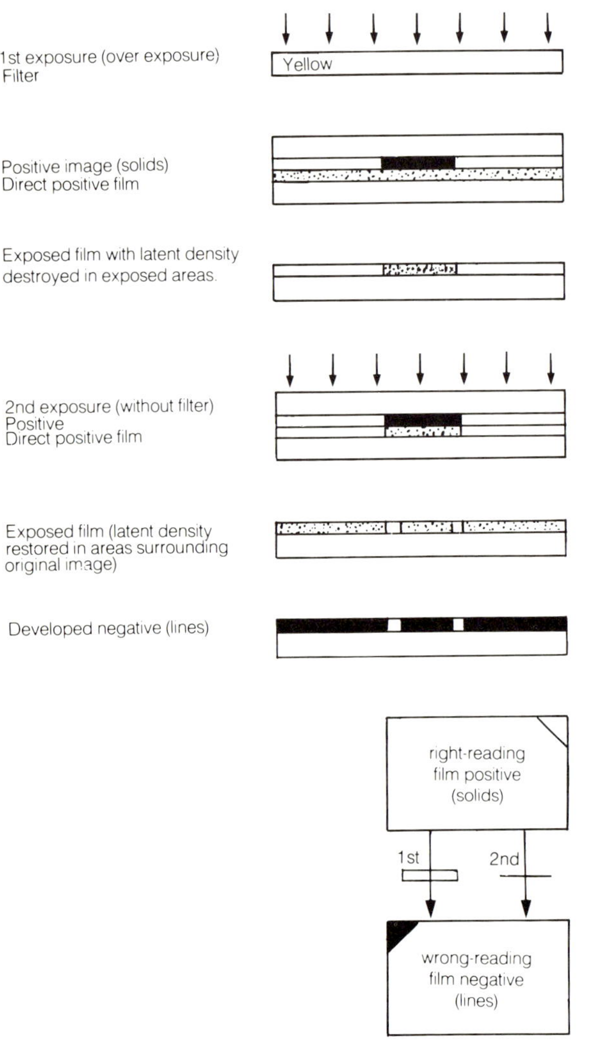

Fig. 7.32 Contact copying arrangement for converting solid areas to lines

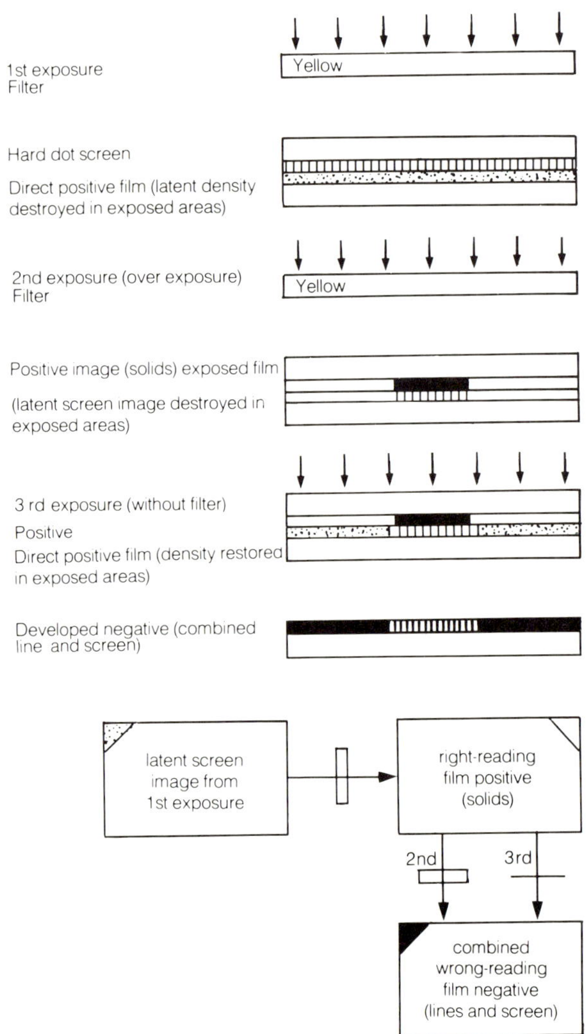

Fig. 7.34 Contact copying arrangement for producing a combined line and screen image

areas. The length of the exposure influences the width of the lines. This first exposure is then followed by a second exposure using ultraviolet light with the filter removed but with the original and film still intact. This second exposure restores the density of the film in the exposed areas surrounding the original image. When the exposed film is developed, the result will be a negative outline image of the solid area of the original. This negative can, in turn, be contacted to positive form, during which the width of the lines can be increased if necessary.

If a combined line and screen image is required (Fig. 7.33) it can be achieved by first exposing direct positive film in contact with a tint screen through a yellow filter, so that the film will contain a latent image of the screen. The subsequent contacting, exposure and developing sequences are the same as for lines, except that the end result will be a combined line and screen image. Figure 7.34 illustrates the copying arrangement.

The technique is used mainly to produce outlines of details which have been generated as solid lines and areas by both manual and automatic drawing techniques, or by the dry transfer (see 1.5) and stick-on techniques.

7.5 Photomechanical vignetting

Photomechanical vignetting is used to accentuate an area without having to completely cover it with a tint or solid colour which can sometimes detract from the overall clarity of other map information. This is achieved by enhancing the boundary line of the area with a band of half-tone dots that

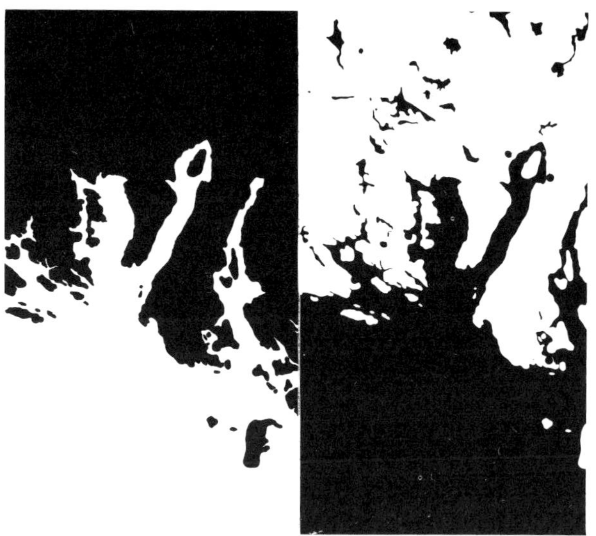

Fig. 7.36 Negative and positive masks of water area

gradually diminish in size away from the line, and blend into the non-print area of the paper (Fig. 7.35). The technique can also be used to enhance line and point symbols.

If, for example, a coastline has to be vignetted, both negative and positive masks of the water area are made (Fig. 7.36). These are assembled and registered in a vacuum printing frame along with the other materials required to make the vignette as follows.

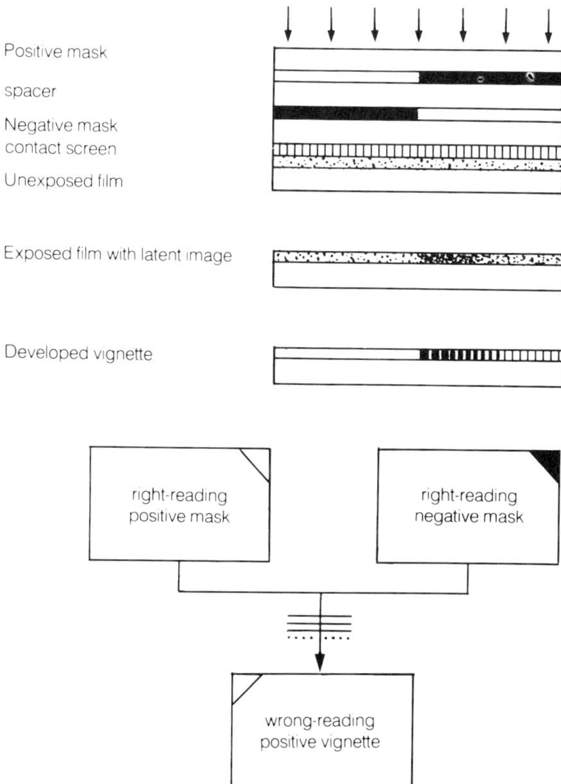

Fig. 7.35 Vignetted area

Fig. 7.37 Exposure arrangement for vignetting

A half-tone contact dot screen is placed over a sheet of unexposed film, then the negative mask of the water area is registered on top of the screen. A spacer, normally opal plastic, plexiglass or several sheets of matt plastic, is placed over the negative mask. The positive mask is then registered over the separator and negative mask (Fig. 7.37). Depending on the required width of the vignette, the spacing between the positive and negative masks may vary from a few tenths of a mm to approximately 10 mm.

When the assembled material is exposed, the light is diffused by the intervening separator and penetrates to the film through the space between the negative and positive masks (Fig. 7.38). The intensity of the light during exposure is greatest at the dividing line and gradually decreases away from the line; so when the film is developed, a screen vignette of diminishing size dots is produced.

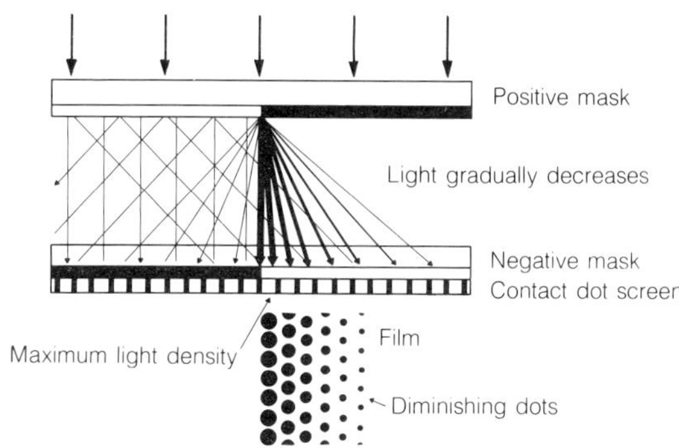

Fig. 7.38 Diffusion of light through intervening separator

7.6 Half-tone to continuous-tone conversion

This technique can be used if a half-tone image has to be 'unscreened' or converted back to its original continuous-tone form. It is especially useful if the original continuous-tone drawing or photograph is unavailable or unsuitable for reproduction purposes. The procedure is to contact or photograph the existing half-tone negative or positive through a diffuser using a continuous-tone film (Fig. 7.39). The scattering of the light through the diffuser during the exposure results in unsharp dots which have a very flat density profile. This causes them to merge in the developing process, thereby restoring the continuous-tone characteristics of the original. The restored image can then be screened again at any required angle.

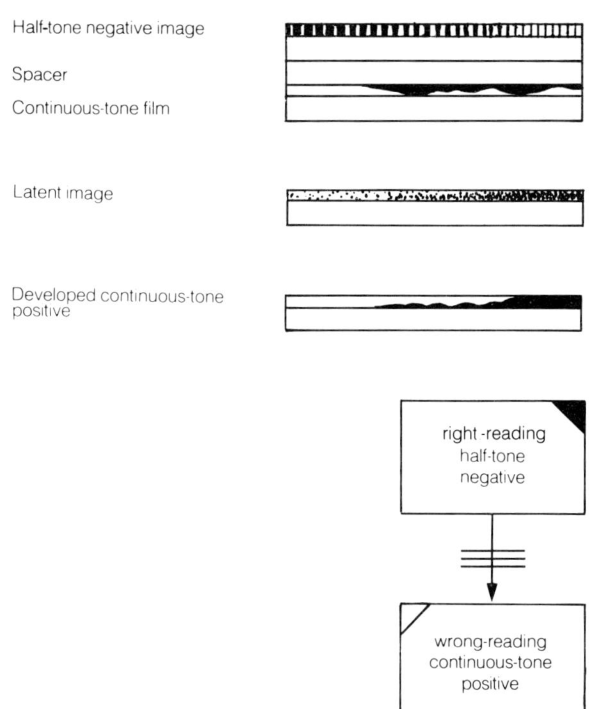

Fig. 7.39 Exposure arrangement for half-tone to continuous-tone conversion

In cartography the technique can also be applied in the following circumstances:

—When the density value of the half-tone dots in a negative or positive of a shaded relief drawing, that was originally prepared for printing in grey (Fig. 7.40), has to be changed for printing in black or some other colour (Fig. 7.41).

—To convert the half-tone image of a monochrome shaded relief drawing into a set of angled half-tone films for multicolour printing. A similar example is the preparation of a half-tone negative for the yellow tone (usually printed on the 'sunny side' slopes of hill shadings) from the half-tone negative that was prepared for printing in grey or black (Fig. 7.42).

—To convert the half-tone screen values of shaded relief or orthophotomap images which have to be photographically enlarged or reduced that would otherwise result in the half-tone dots being either too large or too small.

As it is not possible to print actual continuous-tone images, Figs 7.41 and 7.42, therefore, show only the results obtained following the conversion of the original half-tones into continuous-tone images and subsequence screening for printing in three colours.

Fig. 7.40 Original half-tone negative of shaded relief drawing for printing in grey (screen angle 45°)

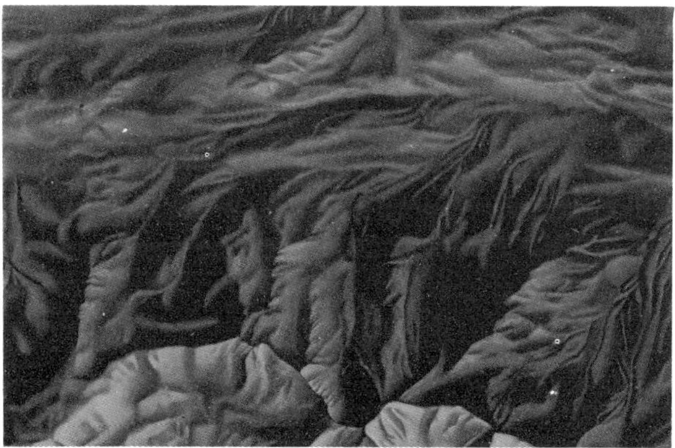

Fig. 7.41 Half-tone negative of shaded relief for printing in blue, after line to continuous-tone conversion and subsequent screening (screen angle 105°)

Fig. 7.42 Half-tone shaded relief negative for printing in yellow, after line to continuous-tone conversion, subsequent screening and reversing to negative form (screen angle 75°)

7.7 Changing contrast for half-tone screening

The magenta and grey half-tone screens have a fixed density range of approximately 1.2. Therefore, if a continuous-tone original does not have the same density range, it is not possible to reproduce precisely the tones of the original. However, it is possible to increase or decrease the overall tonal values of the original before it is screened, so as to provide a copy that has the necessary contrast to give the desired printing results. This can be achieved during the developing process by applying different developing times. In general, the longer the film or photographic print is developed, the higher the contrast will be until it finally reaches a limit.

Fig. 7.43 Results obtained from a normal negative with a full tonal range (reproduced in half-tone)

Using this technique, continuous-tone films or prints of shaded relief drawing and orthophotos can be pre-processed to provide the best results in order, for example, to match the tonal values of other adjoining relief drawings or orthophotomaps.

The photographer should be advised as to the tonal range required, so that he can adjust the exposure and development times accordingly to produce either a normal continuous-tone negative

Fig. 7.44 Results obtained from a flat negative which does not have a full tonal range (reproduced in half-tone)

or print with a full tonal range (Fig. 7.43), a flat or soft negative in which the light areas are veiled and the shadow do not carry a full density (Fig. 7.44), or a contrasty or hard negative with a dropout in the highlight areas (Fig. 7.45).

Fig. 7.45 Results obtained from a high contrast negative with dropout in the highlight areas

7.8 Continuous-tone to line conversion

A continuous-tone image can be photographically converted to a line copy by using a high contrast film, commonly referred to as a lith-type film. When this type of film is exposed and processed in the normal manner, it records only the extreme ends of the grey scale with no tonal gradations in between. This means that the grey scale is virtually split in two with the highlights and middle tones dropped out in one half while the strong shadow tones remain as solids in the other half. The few remaining grey tones in the small transition area between the two can be eliminated in a subsequent stage in the reproduction process. The level at which the grey scale is split can be regulated, to some extent, by varying the exposure and developing times.

The need to convert continuous-tone images into line images in cartography is very limited. The technique, however, can be used to prepare a selective copy of an orthophoto in order to produce a line image for a photomap. Guide copies, with line characteristics, can also be made from continuous tone photographs of printed maps for use as source information when compiling other maps.

7.9 Density slicing

The technique used for density slicing is basically the same as that used for continuous-tone to line

conversion (see 7.8). The only difference is all the tones of the grey scale are divided into several density steps or groups rather than into just two steps. To accomplish this, it is necessary to prepare a low density positive and a high density negative of the original continuous-tone image for use as masks. The high density negative mask is produced from the original continuous-tone image and the low density positive mask is made directly from the negative mask. The function of these two masks is to reduce, or modify, the tonal density of the original image so that the light, medium and dark tones are of similar value.

In the exposure process, both the negative and positive masks are precisely registered over the original continuous-tone image and placed in direct contact with the lith-type film (Fig. 7.46) and exposed. When developed, the film only records the intermediate grey tones of the original in accordance with the amount of light that penetrated the combined densities of the two masks. Depending on the density range of the two masks, selective grey tones or steps in the original can be

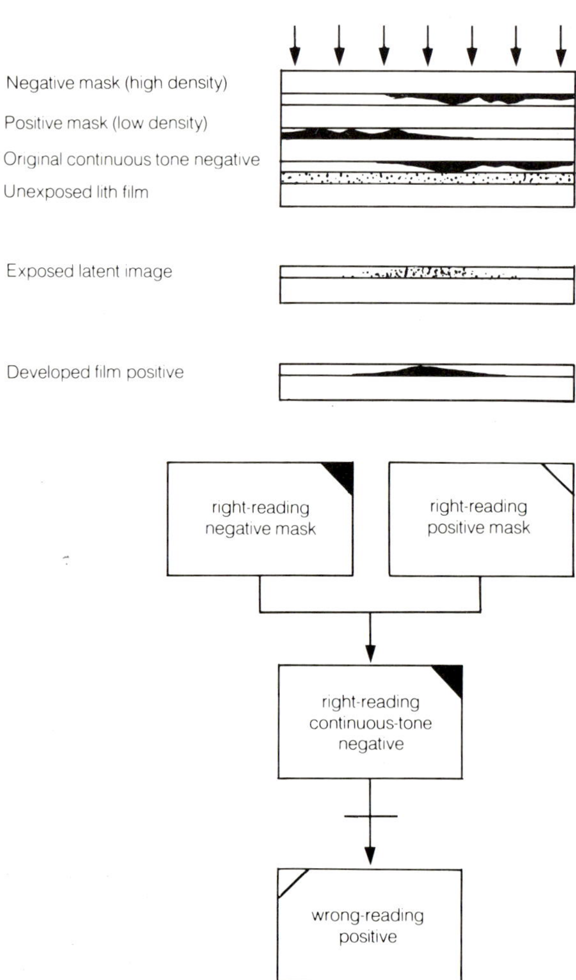

Fig. 7.46 Exposure arrangement for density slicing

extracted. Density slicing has few applications in cartography except, for example, to select image elements of a certain density range from continuous-tone orthophotos, satellite imagery, or shaded relief drawings.

7.10 Edge enhancement (photoline technique)

The edge enhancement, or photoline technique, is a photographic process used to produce an edge or line where there is a sharp change in density in a continuous-tone image.

The technique uses a masking system, in which a high contrast positive mask is made, by contact, from a high contrast continuous-tone negative. Both of these are assembled, in register, with transparent plastic spacers placed in between, along with an unexposed lith type film in a vacuum printing frame (Fig. 7.47). During the exposure, no light penetrates the opaque areas of either the

Fig. 7.48 Line elements of scattered wooded areas from an orthophoto combined with linear detail

negative or positive mask. However, the transparent spacers allow a certain amount of light to spread between the edges of the negative and the positive mask which following development, results in a line image where there were sharp changes in density in the original image (Fig. 7.48). In the areas where the density difference is only very slight, no line image will appear. The width of lines that are created, can be controlled by the thickness of the spacers. Another method which can be used to produce a photoline image is to place the assembled material on a rotating turntable and illuminate it from an oblique angle during exposure (Fig. 7.49). Similarly, the light source can be rotated.

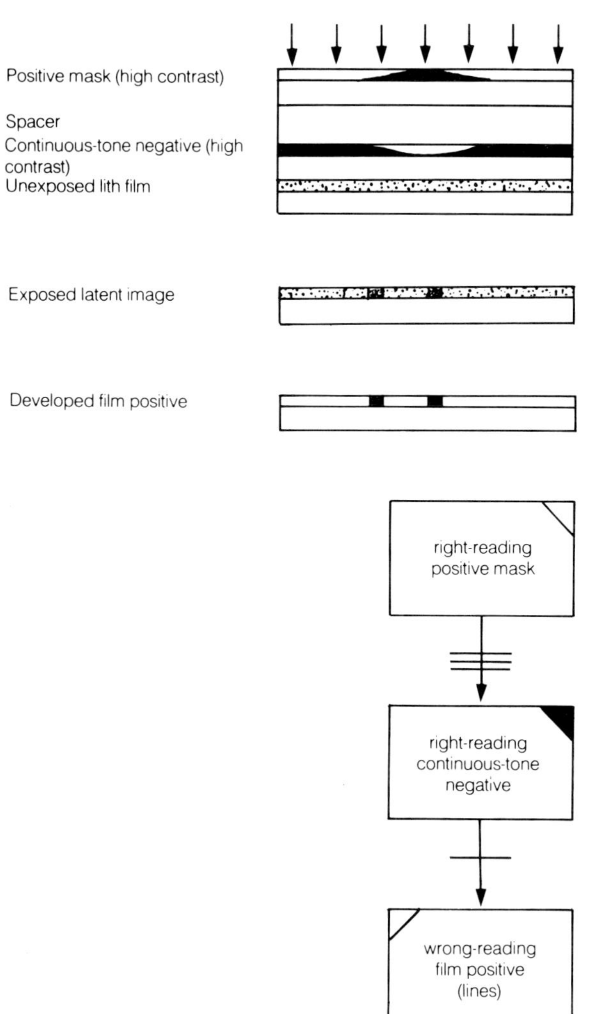

Fig. 7.47 Exposure arrangement for edge enhancement

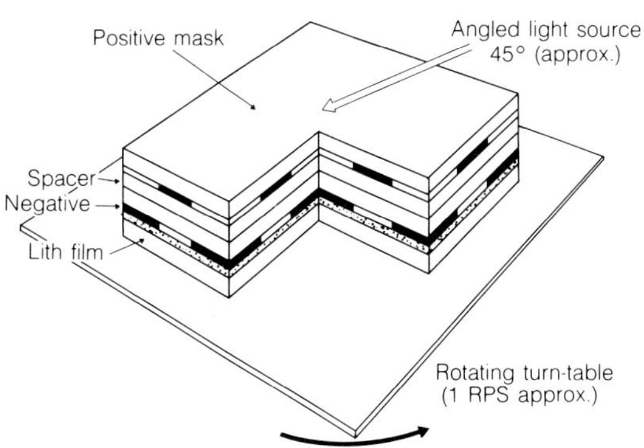

Fig. 7.49 Assembled materials illuminated from an oblique angle and exposed while rotating

Edge enhancement is mainly used in the production of photomaps, orthophotomaps and satellite imagery, in order to enhance the linear structures of features, such as, buildings, wooded and cultivated areas and similar map elements. Unfortunately, the shadows of such features are also recorded.

7.11 Phototone technique

The phototone process is used to convert a continuous-tone image into a fine random dot pattern in either positive or negative form, which can be transferred directly onto a printing plate for printing without the use of a half-tone screen. The sequence of production is as follows.

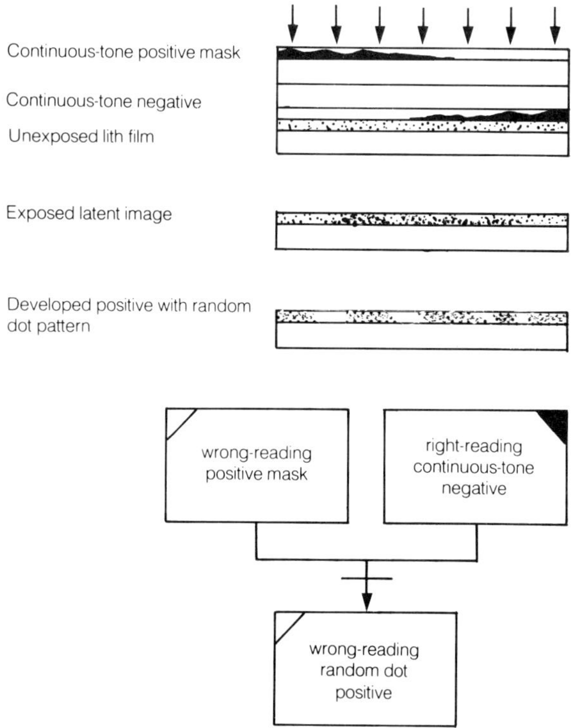

Fig. 7.50 Exposure arrangement for producing a phototone positive

A continuous-tone negative with a limited density range of 1.2 is made by photographing an airphoto or orthophoto. A contact positive mask with a reduced density range of about 70% is then made from the negative. The continuous-tone negative is assembled (Fig. 7.50) in a vacuum printing frame with its emulsion side in contact with the emulsion side of an unexposed lith type film. The positive mask is then registered precisely over the negative with its emulsion side up. When exposed, the amount of light which passes through the assembled materials is controlled by the combined densities of the negative and positive images, which is usually in the 0.3 range. When developed, the film will contain a positive image that has a high density random dot pattern representing the tonal differences of the original. This is known as a phototone image (Fig. 7.51).

If it is necessary to produce a multicolour orthophotomap from a black and white original, then

Fig. 7.51 Phototone image

two contact negatives are made from the phototone positive and one positive is made onto a direct positive film. One of the negatives is made with a normal exposure and contains all of the details of the phototone positive. This negative is used to print the basic map details in one of three selected colours. A second negative has been overexposed, so that only the shadows and darker areas of the phototone image are retained. This negative is used to overprint and accentuate the basic map details using a darker printing colour. The positive, made with a normal exposure, is used to print the highlights of the image using a much lighter colour than that which was used for the basic map details.

The process described to produce a coloured image from a black and white original is basically a combination of the phototone and photoline (see 7.10) techniques. The printed result is commonly referred to as a pictomap, a term derived from photographic image conversion by tonal masking procedures.

The main advantage of the phototone technique is that it eliminates the use of a half-tone screen to break up the tonal values of the original continuous-tone image. It also makes it possible to convert a black and white original to a multicolour image without any tonal discontinuity or moiré patterns.

7.12 Block-out technique (also called hold-out or fat mask technique)

When positioning geographical place and feature names on a map, it is often necessary to place them over line detail. Unfortunately, this can interfere with the legibility of both the name and the feature. To overcome the problem, the lines may be blocked out by manually opaquing them in the areas where the names cross. If the map should contain complex line details and many names, such as on a small scale map of a densely populated area, this can be a very time-consuming operation. However, the same results can also be achieved photomechanically with the use of a block-out mask.

The technique requires that separate name overlays are prepared for each colour component so that the subsequent name negatives do not include any other line details. (This is normally standard practice in most cartographic organizations in order to facilitate production and the maintenance of the map.) The positive block-out mask, on which the line weights of the letters have been increased (see 7.3), is derived from each of the name negatives by the contacting process using diffusing sheets. The positive mask is then combined, in precise registration, with the appropriate line negative to block-out the lines where the names will appear on the printed map. Figure 7.52 illustrates the production sequence.

The use of this technique saves time that would otherwise be used to block-out manually the lines by opaquing. Furthermore, the original line detail remains intact. The technique can also be used to

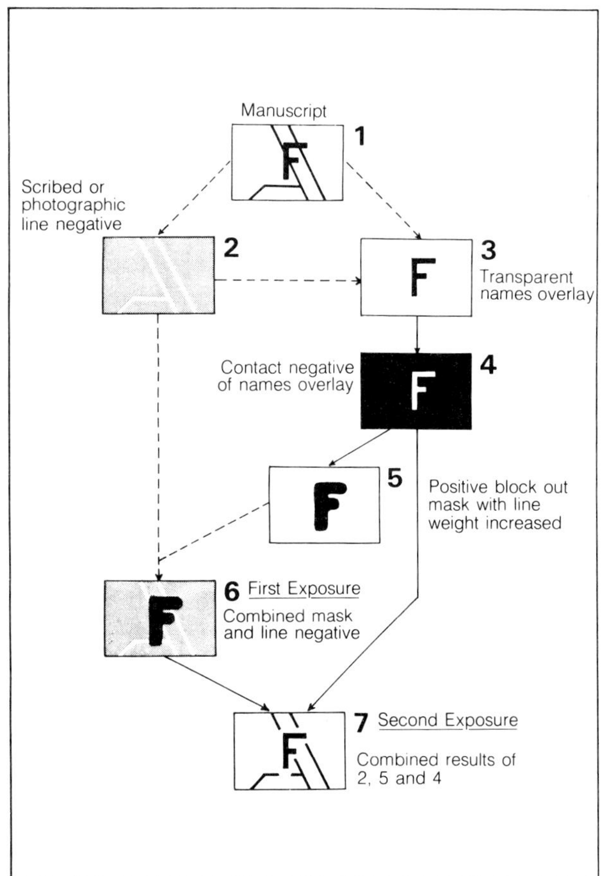

Fig. 7.52 Production sequence for block-out technique

block-out screened areas in which the lettering is to appear in the same colour.

Image Separation

The image separation techniques referred to in this section are used to separate the various image elements of a map which were originally drawn as a single monochrome or coloured image, so as to form individual image components for printing in their own distinctive colours. These processes are commonly known as 'colour separation'.*

The necessity for colour separation lies in the fact that multicolour maps are produced by the sequential overprinting of a number of separate colour components, each depicting a particular map element. This requires that combined line originals be separated into the same number of images to comply with the required number of printing colours.

In most cartographic organizations today, the various line components and colour masks used in the production of a map are produced separately in order to facilitate production and the maintenance of the map. Also, with the use of present day digital technology, one of the principal advantages is that, in an efficiently built-up data base, the various map elements are separated by definition. This allows for the different geographical features, or their attributes, to be addressed directly and manipulated in many different ways as well as separately. This makes it possible to extract and assemble a new digital file of all the features which are to be printed in the same colour (see chapters 2 and 4).

8.1 Manual colour separation of line work

When fair draughted originals, in which the line work for all the printing colours are combined, either as a multicoloured or black line image, have to be manually separated into different colour components the following procedure is normally used:

The combined line original is photographed to produce the same number of negatives (Fig. 8.1) as the printing colours required. On each of these negatives, all the map elements that are not required are deleted with an opaquing solution, leaving only those elements that are to be printed in the same colour (Figs 8.2, 8.3 and 8.4). Although this method saves materials, it is a very laborious

*The term 'colour separation' as used in the graphic arts industry in general, refers to the separation of multicoloured art work into four separate images for printing in yellow, magenta, cyan and black respectively, (see 7.1 and 8.3). Whereas, in cartographic reproduction the range of colours used is often quite different.

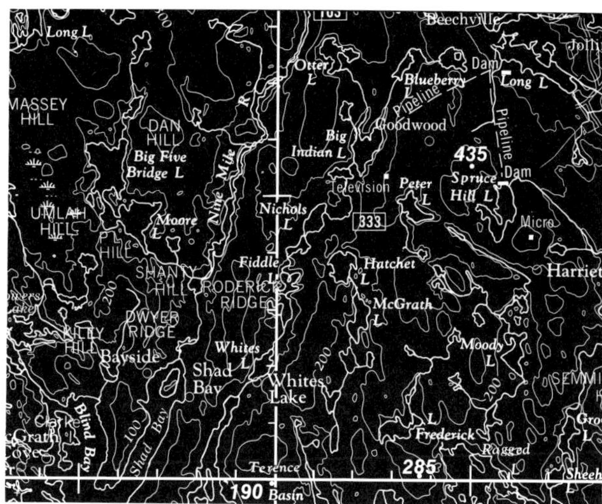

Fig. 8.1 Negative of combined line and type elements of all printing colours

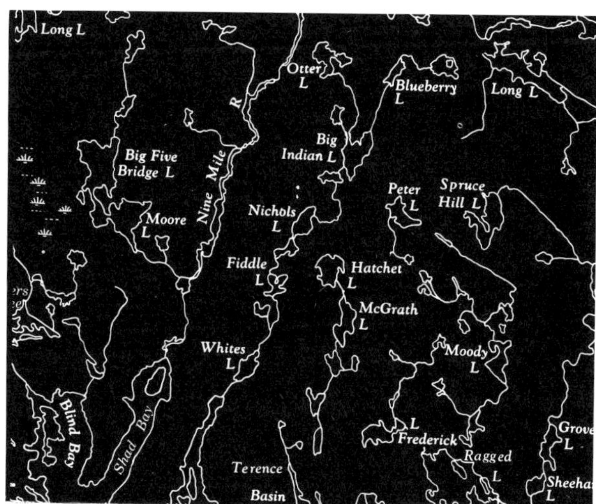

Fig. 8.2 Negative of blue line elements only

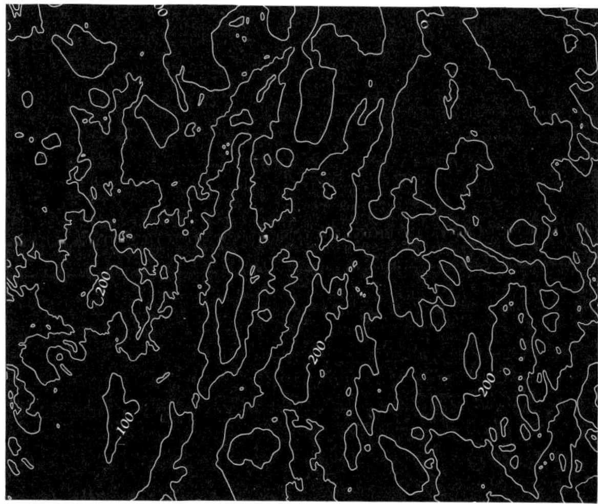

Fig. 8.3 Negative of brown line elements only

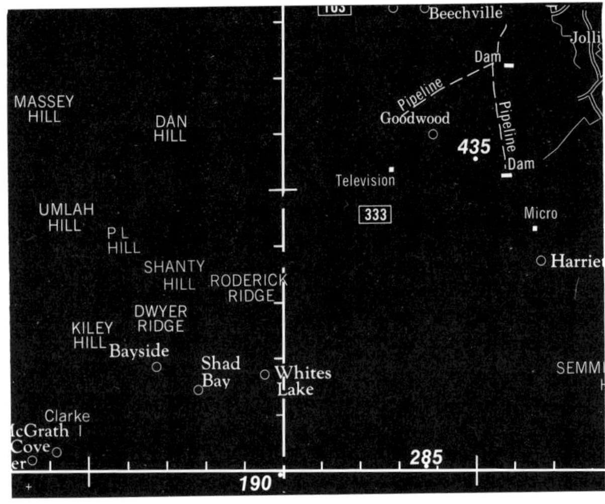

Fig. 8.4 Negative of black line elements only

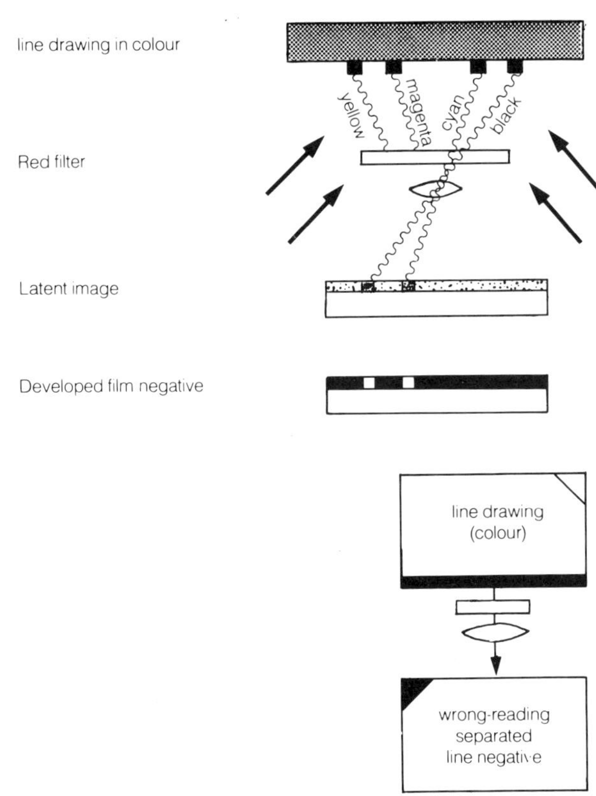

Fig. 8.5 Colour separation of line detail

and time-consuming operation that must be carried out with extreme care and accuracy in order not to opaque and break lines of another colour on any of the negatives. Therefore, this method should only be used for the colour separation of simple maps.

8.2 Photographic colour separation of line work

To reproduce a coloured line original, it is necessary to separate the linear elements of the original into three or more photographic images, each of which are to be printed in a specific colour. This is accomplished by photographing the original through the complementary colour filters (red, blue and green) of the three subtractive primary colours, yellow, magenta and cyan (Fig. 8.5). The colour to be reproduced should be absorbed by its complementary colour filter and the colours that are not required should pass through that filter. Therefore, it is possible to separate the primary colours of the original (Fig. 8.6). Black however, which is the subtractive colour mixture of yellow, magenta and cyan, cannot be filtered out and is

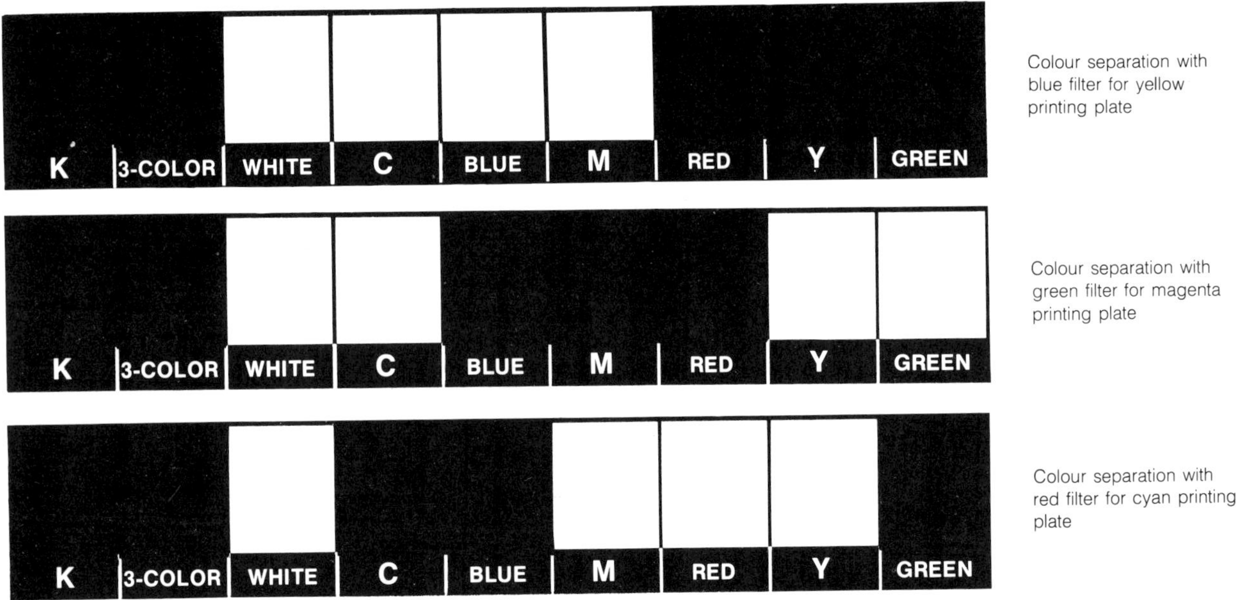

Fig. 8.6 Colour separation of linear elements from the three primary colours

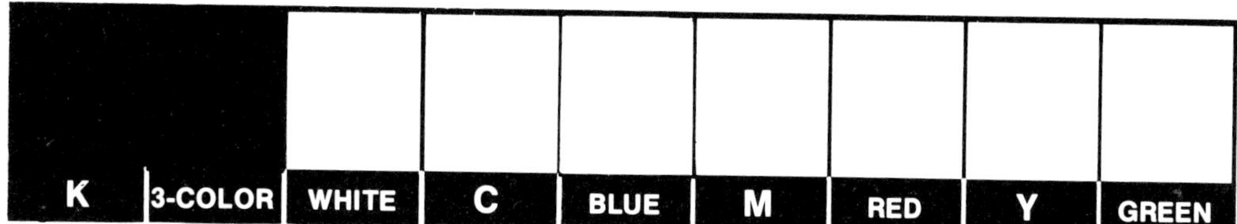

Fig. 8.7 Filtering of the primary colours from black by successive exposures through red, green and blue filters

present in all colour separations. In other words, the primary colours can be successfully separated from black (Fig. 8.7) but black cannot be separated from the primary colours.

In addition to the three subtractive primary colours, most maps contain additional colours which are mixtures of either two or all three of these colours. These are known as secondary and tertiary colours respectively. A secondary colour such as red, for example, can be separated out of a multi-colour original by successive separate exposures through the green and blue complementary colour filters of the two primary colour mixtures, i.e. yellow and magenta (Fig. 8.8). Colour separation of the tertiary colour mixes, such as, brown, olive green, greyish violet and ochre is not always possible photographically. It is, however, a general principle that they can be separated, to some extent, by using a filter of the primary colour that is closest in colour to them, or by using filters of the secondary colour mixes. For example, as ochre is closest in hue to the primary colour yellow it may be separated with a blue filter, and as brown is closest to the secondary colour red, it may be

separated by exposure through a green and blue filter. In either case, some retouching of the negatives may be required to remove some of the lines that are partially visible.

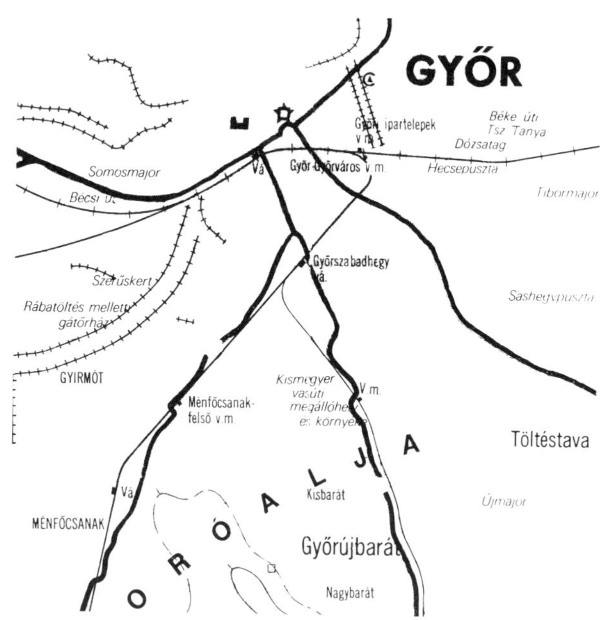

Fig. 8.9 Separated yellow detail

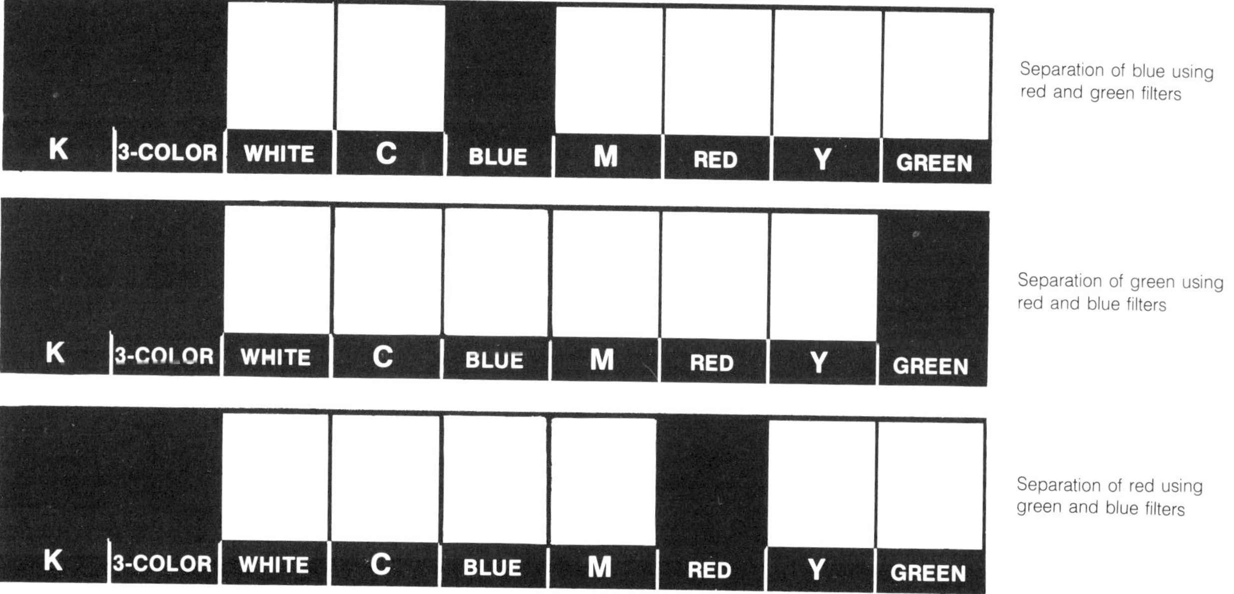

Fig. 8.8 Colour separation of secondary colour mixes

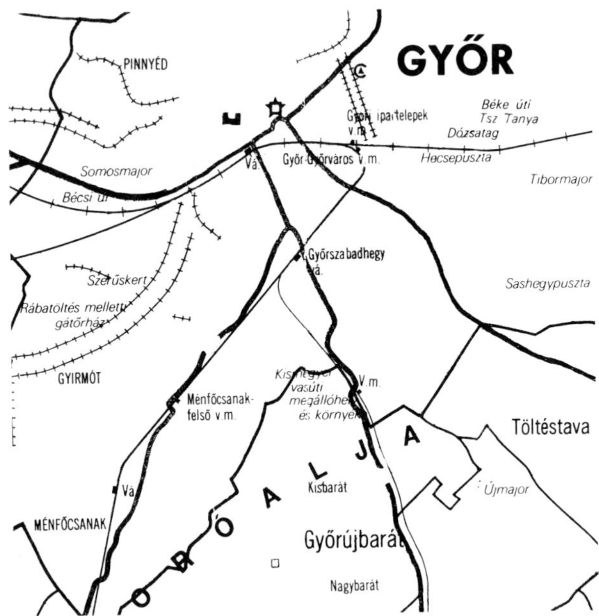

Fig. 8.10 Separated magenta detail

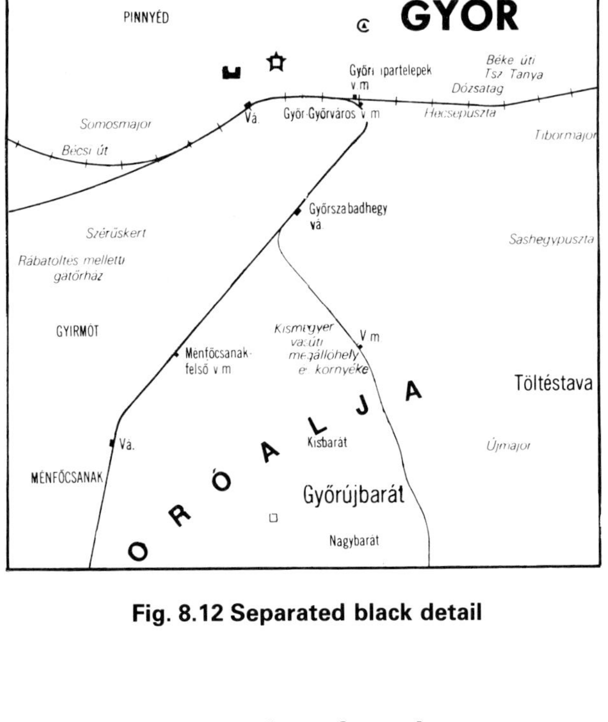

Fig. 8.12 Separated black detail

The quality of the results of colour separation of linear elements is, of course, determined by the quality and character of the original copy. Optimum results can only be achieved by the separation of the primary colours i.e. yellow, magenta and cyan. Black can only be removed from the primary, secondary and tertiary colour mixes by manual opaquing or masking.

Figures 8.9 to 8.12 provide a typical example of the colour separation of linear elements for printing in yellow, magenta, cyan and black.

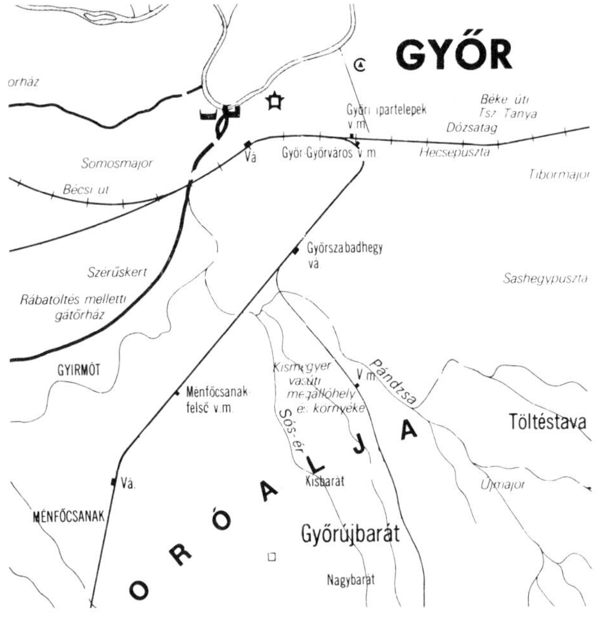

Fig. 8.11 Separated cyan detail

8.3 Colour separation of continuous-tone images by photography

The application of continuous-tone colour separation in cartography is mainly for the reproduction of coloured shaded relief drawings, coloured photo or orthophoto maps, satellite imagery and facsimilies of early maps. As these originals contain many distinctly different tones and shades of colour, it would be impossible to print each separately. Fortunately, most colours can be reproduced by some mixture of the three primary colours, yellow, magenta, cyan along with the addition of some black. (The terms, magenta and cyan, are used in place of process red and blue, because they more accurately describe the actual colours of the printing inks and because it avoids confusion with the blue and red colours of the colour separation filters.) As each of the colours will be printed separately, it is necessary to prepare a set of reproduction negatives or positives, one for each colour.

Unlike the colour separation of line work described in the previous section (8.2), it is necessary to convert the continuous-tones of the originals into a series of dots for printing (see 7.1). This can be achieved by using either the direct or indirect methods of colour separation. With the direct method (Fig. 8.13) the colour separations are made by photographing the opaque or transparent coloured original through the appropriate colour filters and through a glass cross ruled half-tone screen directly onto a high contrast pan-

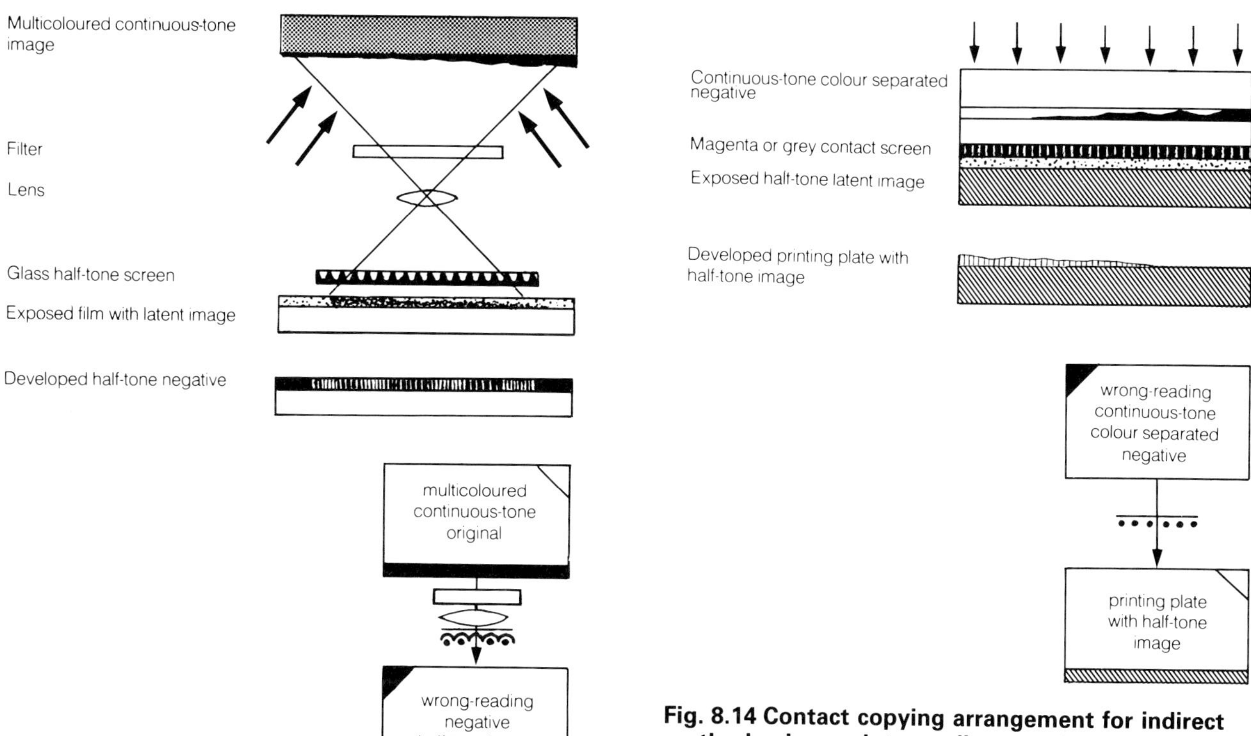

Fig. 8.14 Contact copying arrangement for indirect method using an intermediate continuous tone colour separated negative

Fig. 8.13 Exposure arrangement for direct method of colour separation

chromatic type film. A grey contact screen may be used in place of the glass screen providing the camera is equipped with a vacuum back. The printing plates are made directly from the resulting colour separated half-tone negatives or positives.

With the indirect method (Fig. 8.14), the half-tone negatives are not made directly from the original, but instead, from intermediate continuous-tone separations and transferred to the printing plate, by contacting through a magenta or grey contact screen (see 7.1.3).

The actual colour separation process makes use of the subtractive colour theory, whereby the coloured image is photographed through all the complementary filters of the subtractive primary colours. The complementary colour filters for the primary subtractive colours i.e. yellow, magenta and cyan, are blue, green and red respectively.

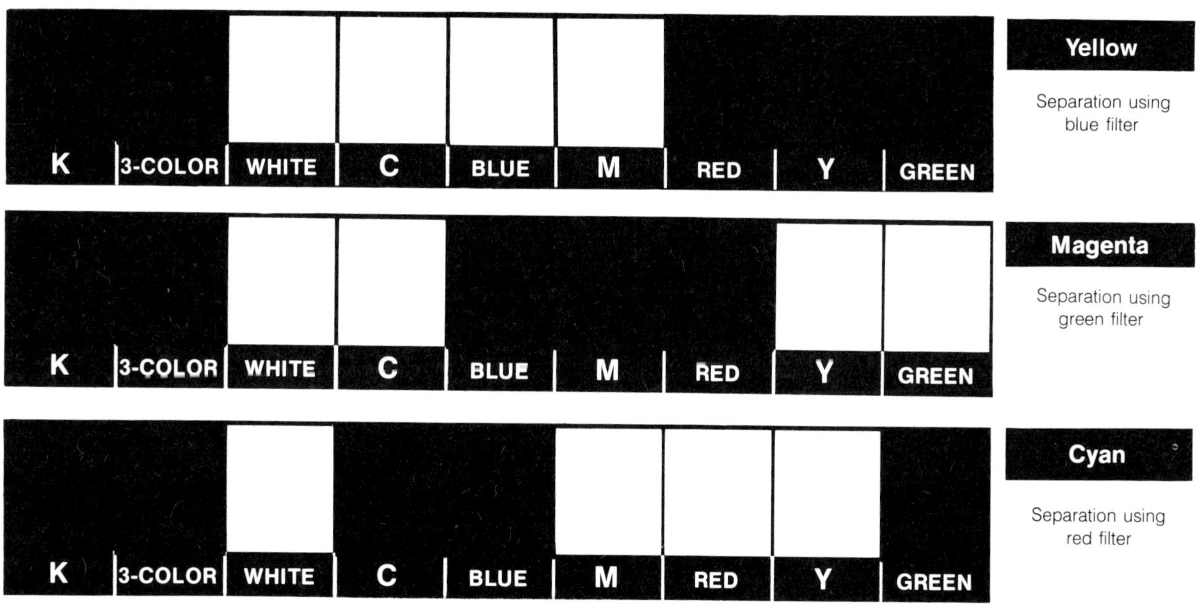

Fig. 8.15 Ideal colour separations

Table 8.1 Separation and non-separation colours using complementary colour filters

Subtractive primary colours	Complementary colour filter	Separation colours with density values equal to black	Non-separation colours with density values equal to white
Cyan	Red	Cyan, Green, Blue, Black	Magenta, Yellow, Red, White
Magenta	Green	Magenta, Red, Blue, Black	Cyan, Yellow, Green, White
Yellow	Blue	Yellow, Red, Green, Black	Magneta, Cyan, Blue, White

In the ideal colour separation of the primary colours and their mixes, e.g. green (cyan and yellow), blue (cyan and magenta) and red (yellow and magenta), their densities should equal that of black when measured through the same complementary colour filter (Fig. 8.15). Table 8.1 shows the separation and non-separation colours of cyan, magenta and yellow.

In practice, the separated colours do not have quite the same density as black, consequently they have less colour intensity and, conversely, the densities of the unseparated colours are greater than that of white (Fig. 8.16). These deficiencies can be traced to the imperfection of the printing inks and the reproduction systems. Chromatic imperfections appear on every colour separation, therefore, true colour reproduction cannot be achieved simply by the use of colour filters alone. This means that to obtain the best possible results the separation negatives or positives should be colour corrected (Fig. 8.17). This is normally done with the use of negative and positive masks.

Negative masks are usually made by photographing the original through the appropriate fil-

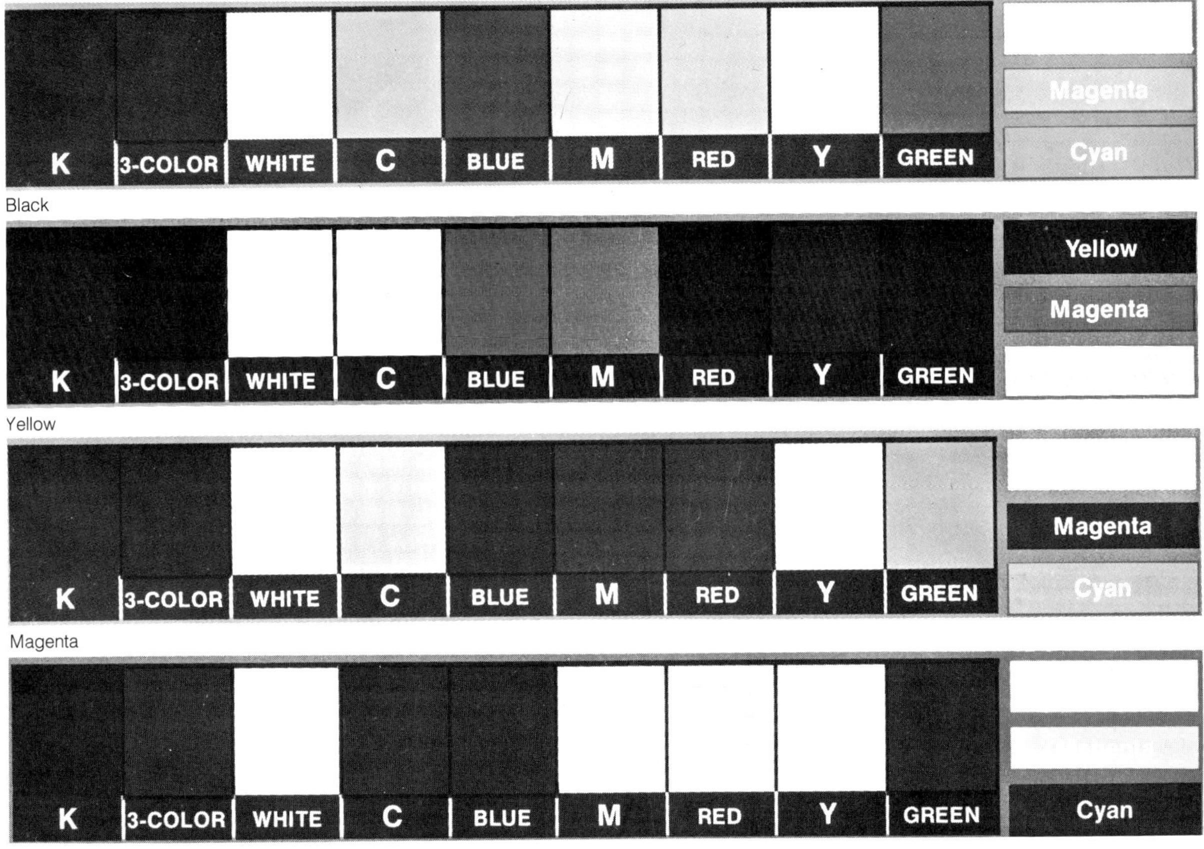

Fig. 8.16 Uncorrected colour separations

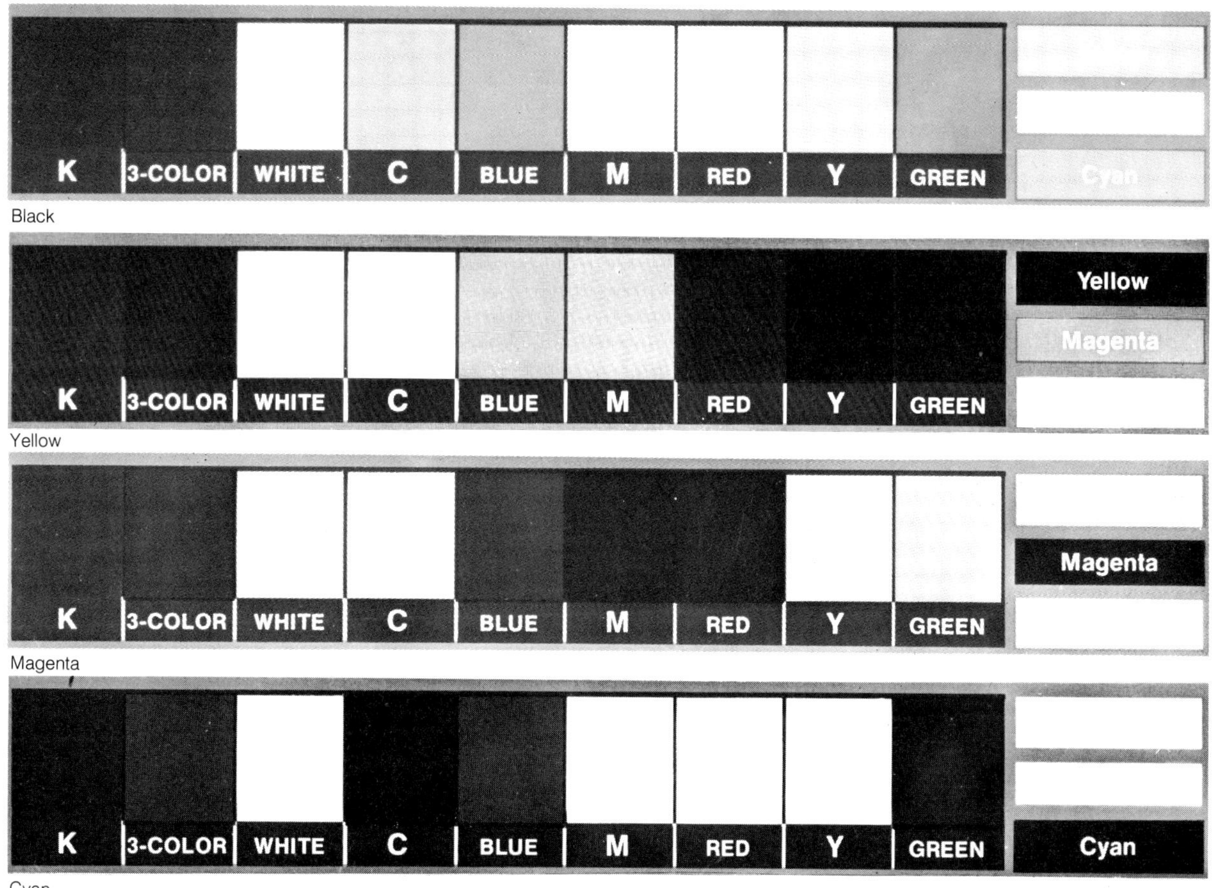

Fig. 8.17 Colour corrected separations

ters before the actual separations are made. These masks are then placed in register over the unexposed film in the back of a camera and the original is then photographed through the selected filter. The amount of light reaching film during exposure is controlled by the choice of filters and the density of the individual masks for each colour separation.

Positive masks are used to colour correct separation negatives by adding to their densities in order to reduce the strength of colour in local areas. They are normally made directly from the uncorrected colour separated negatives. They may also be made from differently filtered negatives in order to vary the amount of colour correction. For example, a positive mask contacted from a cyan

Fig. 8.18 Yellow

Fig. 8.19 Magenta

Fig. 8.20 Cyan   **Fig. 8.21 Black**

filtered negative may be used to correct a magenta filtered negative. The positive mask is assembled, in register, with the negative to be colour corrected and exposed.

While it is possible to come close to representing the true colours of the original copy by colour correcting the separation negatives, it is not possible to match them exactly because of the inadequacies in the printing inks used and the variables in the reproduction sequences involved.

Figures 8.18 to 8.21 provide examples of the four colour half-tone separations of a water-coloured map original.

The colour separation of continuous-tone images, is also done with the use of electronic scanning equipment as described in item 7.1.2.

Image Combinations

In the previous chapter on image separation, the main objective was to prepare a separate reproduction negative or positive for each printing colour by manually opaquing or photographically colour separating the various colour components for the map. However, for ease of production and the future updating of the map information, most colour components are originally divided up into a number of separate elements, which when combined, form a single image for printing in a specific colour. For example, the blue printing image may be made up by combining separately prepared elements, such as, the shorelines and river network, bathymetric contours, solid or screened tints for the lakes and, the lettering for the feature names.

Such combinations may have to be made at various stages during the reproduction process, whether it be to produce a more complete compilation base, to produce a colour-proof or to make the final printing plate. Some of the types of images that are likely to be combined include: line detail, solid and screen tints, lettering, point symbols, diagrams, line and half-tone images and, photoline and phototone images. It is, of course, virtually impossible to list all the possible combinations. Therefore, the following subjects deal only with a few of the more basic methods.

9.1 Combination of negatives by contact printing

The contact printing process (see 5.1 and 5.1.1) is used to combine two or more negative images in order to form a single positive image. During the process, it is essential to obtain perfect register between the separate negative components, therefore, the whole process depends on the use of a very precise punch register system (see 3.2). As the objective is to produce a wrong-reading positive that can be used directly to make a printing plate, the original negatives should be in right-reading form. Also involved in the contacting sequences are the results of many different sub-processes and by-products, such as, scribed originals, type and symbol overlays, colour masks, block-out masks, etc.

When making image combinations, normally the line negative, which contains the main framework of a particular colour plate, is contacted first and, prior to the film being developed, the other line negatives, masks and screens etc., are registered and sequentially exposed, usually in order of their importance to registration. During exposure, it is

extremely important that the image on the emulsion side of the negative is in direct contact with the light-sensitive emulsion side of the lith or line type film that is used. This is especially important when combining dot or line screens. Figure 9.1 illustrates the production sequences for making a combined positive component for printing in blue. The number of separate negative components that are used will vary according to the complexity of the detail. The same procedure, of course, is used to produce combined positives for each additional colour.

The contacting and developing procedures used are quite economic in view of the time involved, but they are critical insofar as the end result can only be inspected after the combined film has been developed. If, for example, one mask was left out or one of the components was registered in reverse, the whole process must be repeated. It is, therefore, important to establish a detailed plan showing the precise contacting sequences prior to combining the various individual components.

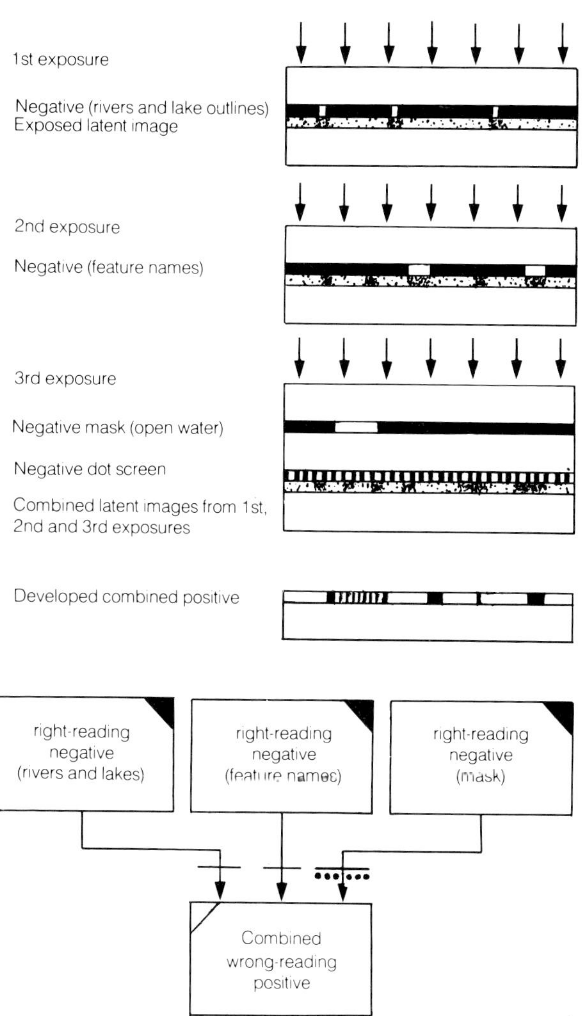

Fig. 9.1 Contact copying arrangement to produce a combined positive image

9.2 Combination of positives and negatives

It is not unusual, when image combinations are necessary, that some of the map components, such as the lettering overlays, will be in positive form, whereas, other components, such as the scribed details, will be in negative form. Normally it would be necessary, for example, to convert the positives to negatives so that all the components are in the same form in order that two or more can be combined to form a single positive. However, by using direct positive film (see 5.1.2), it is possible to combine both a positive and a negative on the same emulsion.

This is accomplished by means of a double exposure. First, the positive image is exposed in a vacuum printing frame through a yellow filter onto the direct positive film. The filter is then removed and the positive is replaced with the negative and the second exposure is made to the unfiltered light source. When developed, the film will contain both the positive and negative images in positive form (Fig. 9.2).

A variation of this procedure is to scribe the line details on a yellow scribe coat material. Type and symbols may then be added in positive form directly to the scribing surface of the material. A combined positive can then be made on direct positive film as the yellow scribe coat acts as a filter (Fig. 9.3).

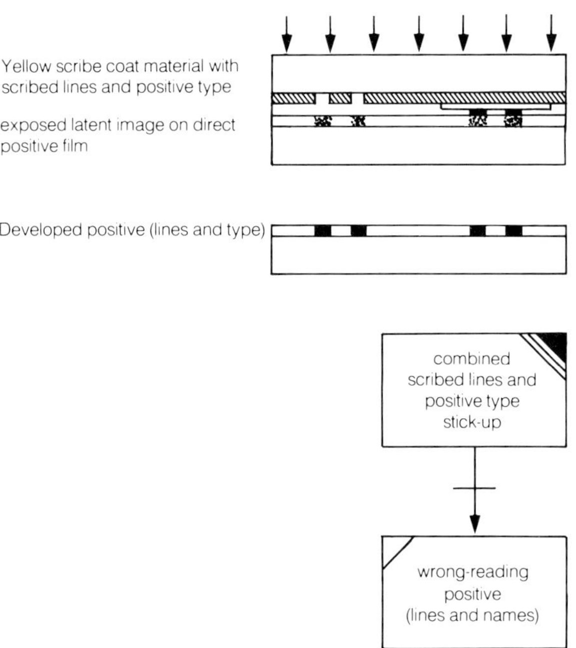

Fig. 9.3 Exposure arrangement for producing a positive from a combined scribed image and type stick-up on yellow scribe coat material

9.3 Photomechanical colour-proofing

Colour-proofs are often required at different stages of map production in order to evaluate the map content in terms of accuracy, quality and colour, and to ensure that the various map components are complete and in perfect registration with each other.

The two main methods of colour-proofing are: press-proofing on paper with printing inks and photomechanical proofing using light-sensitive colour coatings on plastic base materials. Press-proofing is ideal, especially for assessing colour fidelity and compatibility. However, this method requires a special proofing press or the use of a regular offset printing press which is both time consuming and costly, consequently, press-proofs are seldom used today. Instead, most cartographic organizations produce colour-proofs by photo-

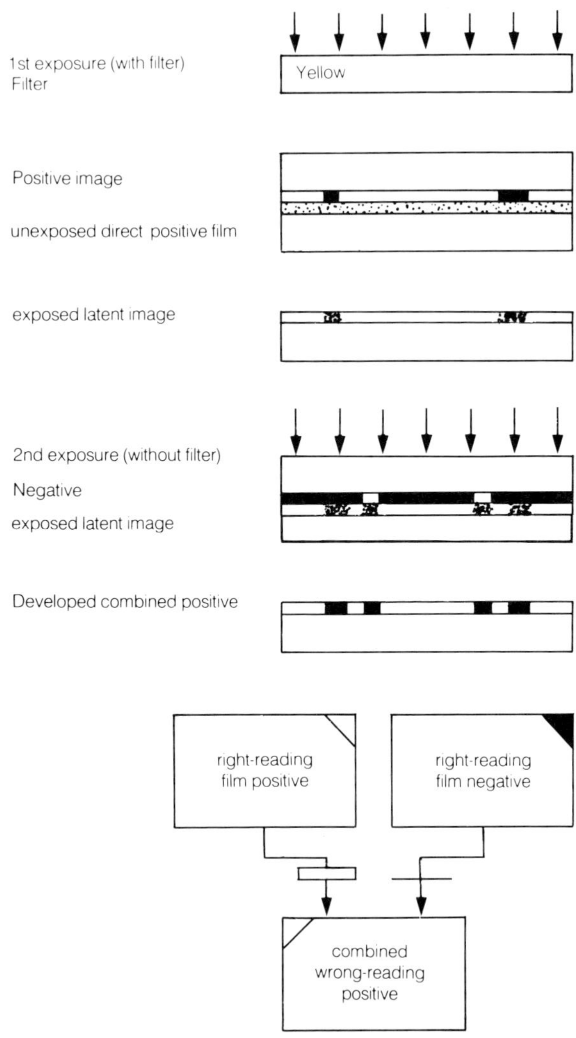

Fig. 9.2 First and second exposure arrangements for producing a combined positive from separate positive and negative images

mechanical means, either by superimposing methods whereby the different colour components are combined on the same surface, or by overlay methods where each colour is on a separate sheet of transparent film which can be overlayed, in register, on a light table for viewing (see 9.4).

There are a number of different positive and negative working superimposing systems available commercially. Some are tied to multicolour reproduction using the standard four-colour process i.e. of yellow, magenta, cyan and black, while others provide a wider range of additional colours. The flexibility to mix dyes and pigments provided by the manufacturer to produce a particular colour is a great advantage in cartography.

Colour-proofs are usually made on translucent or opaque white plastics that are coated with a light-sensitive emulsion for a particular colour, either by wiping the emulsion onto the plastic or by applying it with a whirler. When the coating is dry, the sheet is exposed in a vacuum printing frame along with the appropriate colour component, to an ultraviolet light source (Fig. 9.4). The exposed image is then developed with

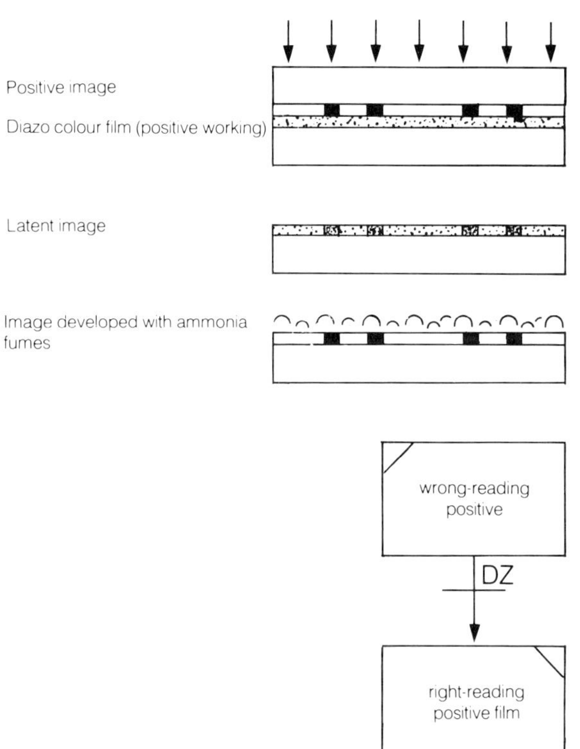

Fig. 9.5 Exposure arrangement using diazo colour film (positive working) for colour-proofing, procedure is repeated onto each separated sheet of film

water which removes the unexposed soluble coating leaving the exposed light-hardened colour image. This procedure must be repeated for each additional colour.

9.4 Combination of colours using overlay methods

With the overlay systems of combining colours, each colour is presented on a separate presensitized sheet of film. Both diazo and non-diazo type films, that are either positive or negative working, are

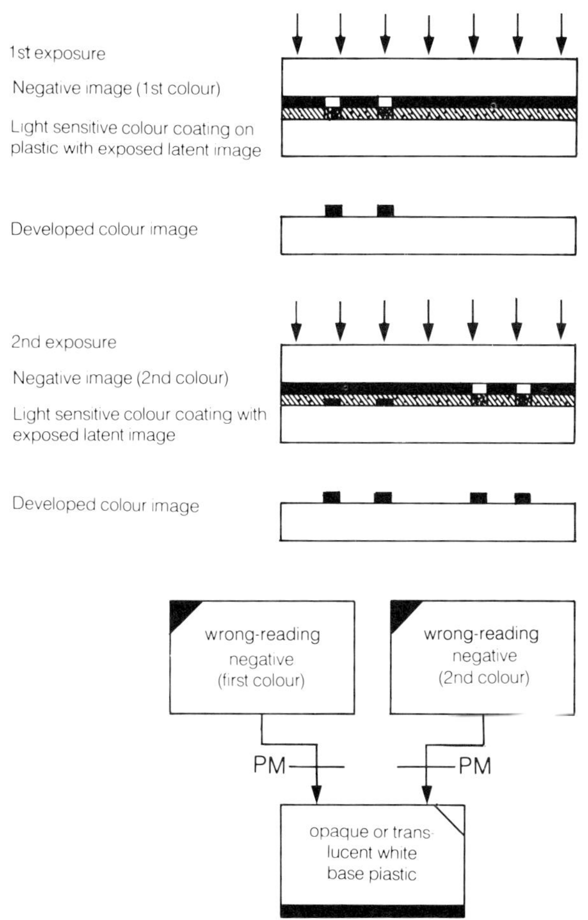

Fig. 9.4 Exposure arrangement for superimposure method of colour proofing (procedure is repeated for the required number of colours)

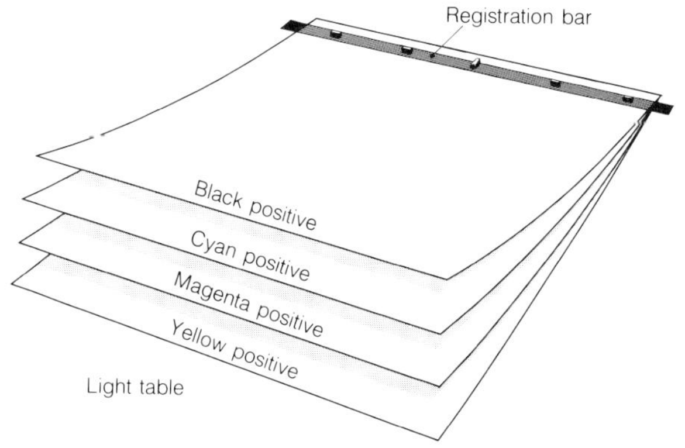

Fig. 9.6 Developed colour films over-layed for viewing

available commercially for this purpose. These films, however, can never match the actual colours of the inks used to print maps. Therefore, the film that will give results closest to a specific printing colour is selected for proofing.

Each film is exposed in a vacuum printing frame together with its own colour component (Fig. 9.5), e.g., the yellow film is exposed with the yellow image component, the magenta film with the magenta component, and so on. The exposed image is developed either by moistening the emulsion surface of the film with a special developer or by exposing the film in a processer to warm ammonia fumes, depending on the type of commercial product that is used. The developed sheets can be overlayed, in register, on a light table for viewing (Fig. 9.6).

The value of such overlay systems in cartography is limited mainly because of the reflection of light through several layers of film and its effect on colour. Besides, for checking purposes, it is preferable to have the various colour components of the map superimposed on the same surface (see 9.3).

Printing Processes

The process of producing multiple copies or reproductions is traditionally associated with a printing press, but any reprographic systems, such as, photographic or electrostatic systems that are capable of producing duplicate copies from a single original may also be described as printing processes.

There are basically four major printing processes: the planographic or offset lithographic process, in which the image and non-image areas are on the same plane; the relief or letterpress process in which the printing areas are raised above the non-printing areas; the intaglio process in which the image is engraved or etched below the surface of the printing plate, and the stencil or silkscreen process in which the ink is pressed through a stencil on a finely woven fabric screen to the paper or some other type of surface.

10.1 Lithographic offset printing

Today, the overwhelming majority of cartographic organizations use the lithographic offset method of indirect printing employing presses that are capable of printing one, two, four or more colours simultaneously at a high rate of speed.

The basic printing unit of a rotary offset press consists of three cylinders, namely plate, printing and impression (Fig 10.1). The printing plate, which carries a printing image that will accept ink but repel water and a non-printing area that will accept water but repel ink, is attached to the plate cylinder. On revolving, it is first dampened by the water rollers and then inked by one of the many inking rollers. The large number of water and inking rollers are necessary to ensure that the water and ink are evenly distributed along the length of the last roller. This is ensured by some of the rollers having a horizontal oscillating movement. After being inked the plate comes in contact with the

printing cylinder leaving its image on the rubber blanket which, in turn, offsets onto the paper carried by the impression cylinder.

The paper is fed into the machine (Fig. 10.2) by means of an automatic feeding device which separates the sheets and causes them to flow along a feed board where they are positioned against the gripper and side lay stops before going into the press. The grippers on the revolving impression cylinder catch the paper from the feed board and hold it in position until it is touched by the rubber blanket carrying the impression from the plate. After the impression has been made on the paper, grippers on the delivery chain remove the paper from the impression cylinder and carry it to the delivery pile.

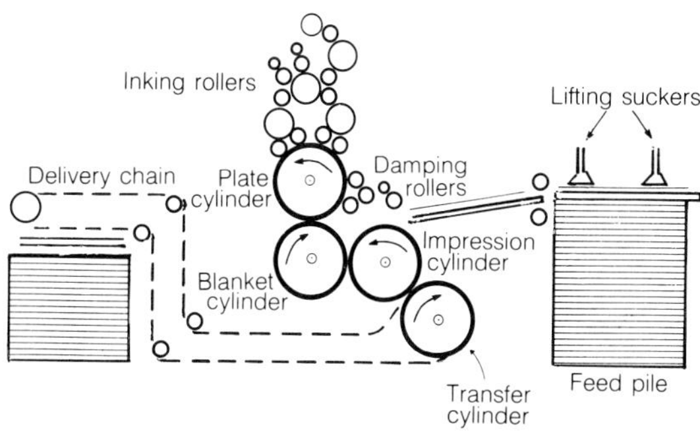

Fig. 10.2 Layout of a single colour lithographic offset printing press

The quality of the final printed product is dependent upon the quality of the printing plate which also has a great influence on the number of impressions which can be made from the plate (see 5.10). Most printing plates today are purchased commercially and can be obtained either precoated with a light-sensitive diazo compound or pre-grained but uncoated, in which case the light-sensitive coating must be applied by hand or in a mechanical roller-type coater (Fig. 10.3).

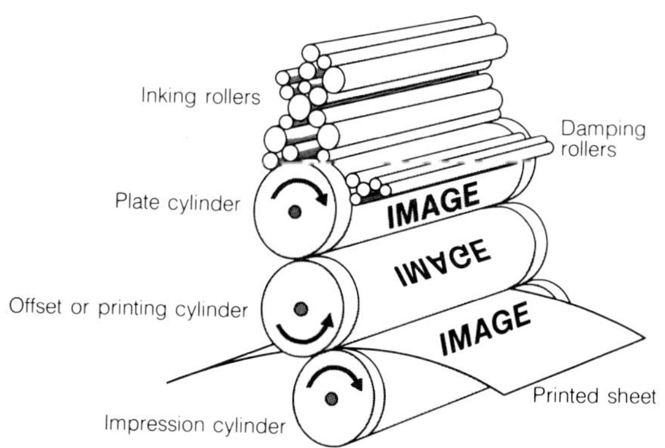

Fig. 10.1 Basic offset printing unit

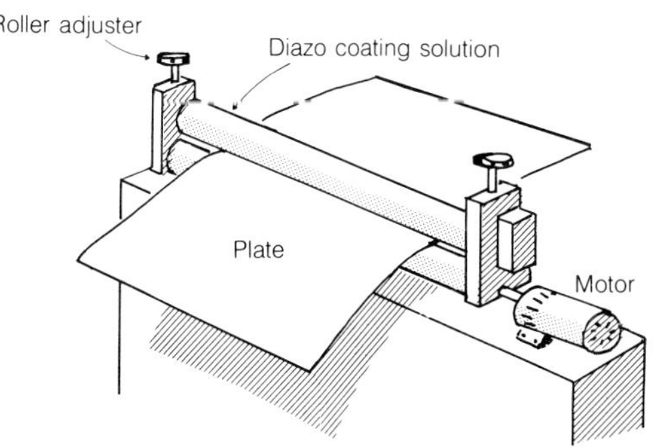

Fig. 10.3 Mechanical plate coater

10.2 Letterpress or relief printing

In relief printing, the printing image is raised on the printing plate. The plate itself can either be flat or cylindrical (Fig 10.4) depending on the type of printing press that is used.

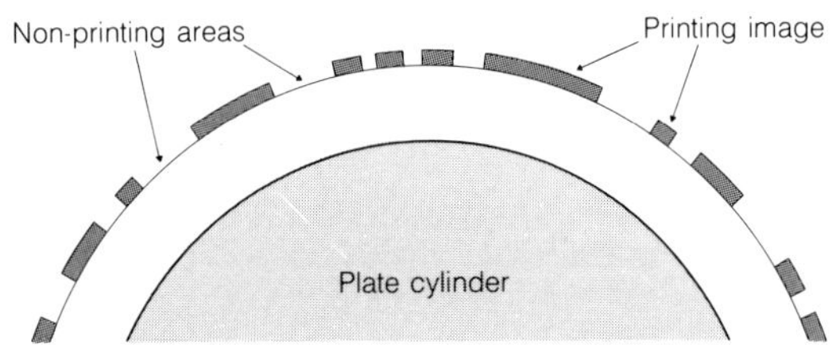

Fig. 10.4 Relief printing with the printing image raised above surface of plate

During the printing process, ink is applied to the raised printing surface and the paper is then pressed against the plate to transfer the image to the paper. If the image is transferred directly from the plate to the paper, the image on the plate must be in reverse or wrong-reading in order that it will be right-reading when transferred to the paper. If the image is transferred indirectly by offset printing, then the image on the plate must be right-reading.

Type can be set by hand or machine. By hand, each letter is selected from a tray of type characters to compose the names and text. By machine, the typesetter presses the letters on the keyboard of a machine that casts the letters in hot metal (see 1.7).

Letterpress printing is extensively used to print books, newspapers, magazines and similar products. It is also used to print small maps where they accompany textual information in books. It can also be used to produce tactual maps for the visually impaired by embossing the braille characters along with the map detail so they are raised in relief on the surface of the paper or plastic material.

10.3 Intaglio printing

Intaglio printing is the opposite to relief printing as the printing image is engraved or etched into the surface of the plate (Fig. 10.5). During the printing process (Fig. 10.6), the entire plate is covered with ink which is then mechanically wiped clean leaving the recessed image filled with the ink. The image is then transferred to the paper by pressing the paper against the plate.

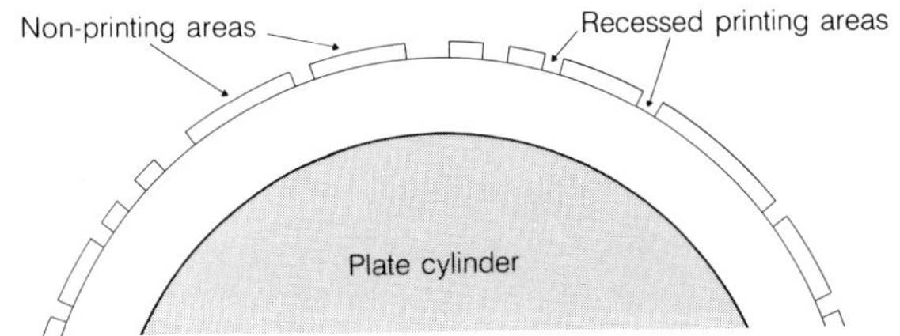

Fig. 10.5 Intaglio printing with recessed printing image

Gravure printing, for example, is an intaglio process which is used to print multicolour illustrations without the use of half-tone screens which are necessary in the lithographic and letterpress printing processes in order to reproduce continuous-tone originals. The variations in the densities of the tones are achieved by the variations in the thickness of the ink in the cells of the gravure plate which is transferred to the paper during printing.

Gravure printing, because of its tonal characteristics, is not generally used to print maps.

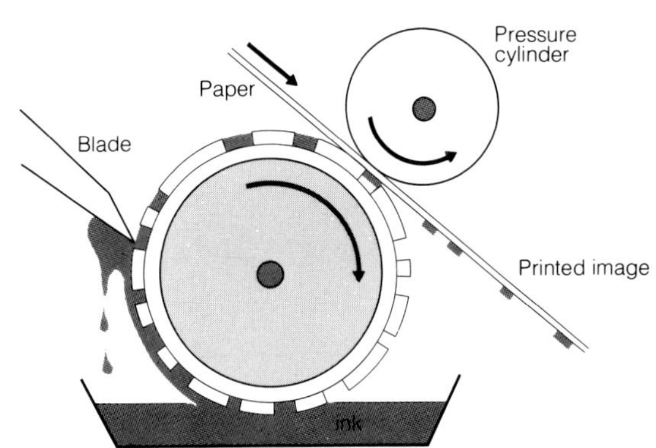

Fig. 10.6 In gravure printing, the image is transferred direct from the plate to the paper

10.4 Silkscreen printing

The silkscreen process uses stencils which can be cut by hand from paper or a special lacquer film and attached to the surface of a fabric or metal mesh screen; or painted by hand directly onto the screen using an oil-based masking solution, or photographically prepared on a special photo-stencil film (see 5.10.1) which is then transferred onto the screen.

For cartographic purposes, however, stencils are more often produced by photomechanical means by coating the screen with a light-sensitive dichromated colloid on which a positive original on transparent film or plastic is exposed and developed to form a stencil. In all cases, the non-image areas are blocked-out on the screen.

Printing is accomplished by forcing the ink through the open image areas of the stencil and screen with a rubber squeegee blade onto the paper or some other type of material (Fig. 10.7). This can be achieved by hand or with a silkscreen printing press. Prints that are made with more than one colour require a separate stencil for each of the colours used. Each colour is allowed to dry before the next colour is registered and printed.

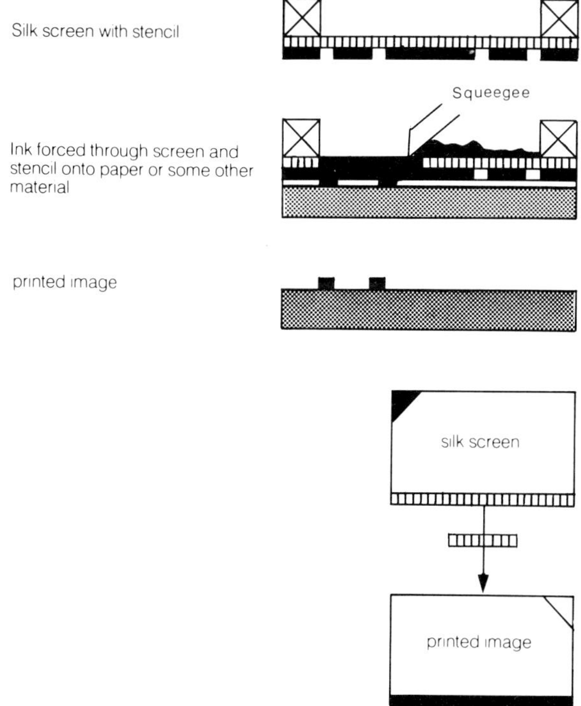

Fig. 10.7 In silkscreen printing the ink is forced through the image stencil and screen onto paper or some other type of material

The silkscreen process can be used to print both line and half-tone images including maps on almost any type of surface, such as paper or board of any size and thickness, plastics, wood, glass, cloth and similar materials.

Although cartography is primarily concerned with the offset lithographic printing process (see section 10.1), there are, however, several other methods of printing, which although not generally used to print maps, the cartographer should, nevertheless, be aware of them. Some of these processes are briefly described in the following sections.

10.5 Collotype (photogelatin) printing

This process is similar to lithography except that it employs a bichromated gelatin image as the printing surface on which detail and tone values depend on the degree to which the printing and non-printing areas of the plate have been made ink-receptive and water-repellent.

A printing plate, usually aluminium, is coated with a light-sensitive layer of bichromated gelatin (Fig. 10.8) and exposed in a vacuum printing frame along with a negative. During exposure, the light hardens the gelatin coating according to the tonal densities of the negative. In other words, the highlight or dark areas of the negative are only slightly hardened, whereas, the shadow or clear areas are hardened to a much greater extent.

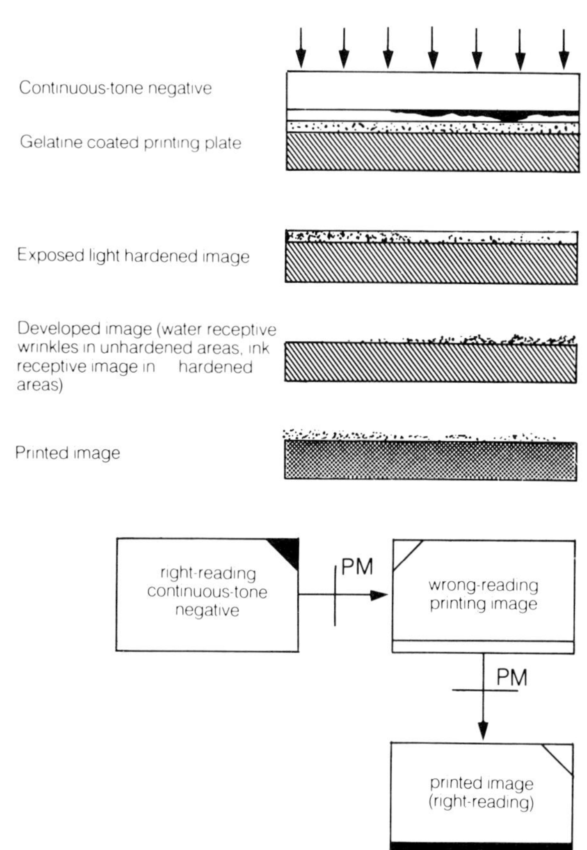

Fig. 10.8 Exposure arrangement for making a photogelatin or collotype printing plate for direct printing

In the developing process, the unhardened gelatin areas become saturated with water, which causes the gelatin to wrinkle leaving a random pattern of irregularly shaped minute points which remain moist. The light hardened image areas resist the water and therefore do not wrinkle.

During the printing process, the moist water-receptive points on the plate repel the ink and do not print, whereas, the slightly lower and dryer image areas accept the ink and therefore print. Because of the irregular pattern of spots (Fig. 10.9) on the plate, the process does not require the

Fig. 10.9 Collotype random dot patterns showing the characteristic wrinkles derived from the gelatin grains

use of a half-tone screen in order to reproduce a continuous-tone image.

The printing image is very delicate, consequently, the number of copies that can be printed from the plate is very limited. The process is mainly used as a direct printing process for fine art reproduction in comparatively small quantities.

10.6 Electrostatic printing

In electrostatic printing processes, an image is produced by the action of light, or infrared radiation, on photo-sensitive semi-conductive surfaces which have been charged by static electricity.

There are two systems in general use: A direct image transfer system in which pigments contained in a liquid toner are attracted by the exposed parts of the image and fixed to the surface of the paper, and an indirect system in which fine powder pigments are transferred from an intermediate carrier to the surface of the paper.

In the direct process (Fig. 10.10), the photo-responsive surface consists of a zinc oxide photo-conductor contained in a silicon resin binder which is coated onto the supporting base material that can be either paper or metal. The coated material is passed under a corona screen and given a negative electrical charge. The sheet is then exposed to the original image during which the non-image areas that are exposed to the light loose their charge leaving a latent electrostatic image on the sheet. The image is made visible with a liquid toner which deposits fine pigment particles, that have been charged with an opposite (positive) electrical charge, and are attracted to the latent image and permanently fixed.

Fig. 10.10 Direct electrostatic process

In the indirect process (Fig. 10.11), the image support is coated with a thin layer of selenium. The plate is passed under the charging screen which applies a positive change to its surface. The image is then projected through a lens and exposed to light which causes the electrostatic charge to be removed in the exposed areas but leaving a charged latent image in the unexposed areas.

During the developing process, a negatively charged fine powder is spread over the plate and adheres to the positive charged latent image, thus providing a permanent visible image. This image is then transferred to another surface which can be accomplished in several different ways. For example, a sheet of plain unsensitized paper can be brought in contact with the charged plate and given a positive charge which, in turn, attracts the powder image from the plate to the paper. This image is then fused to the paper by the application of heat.

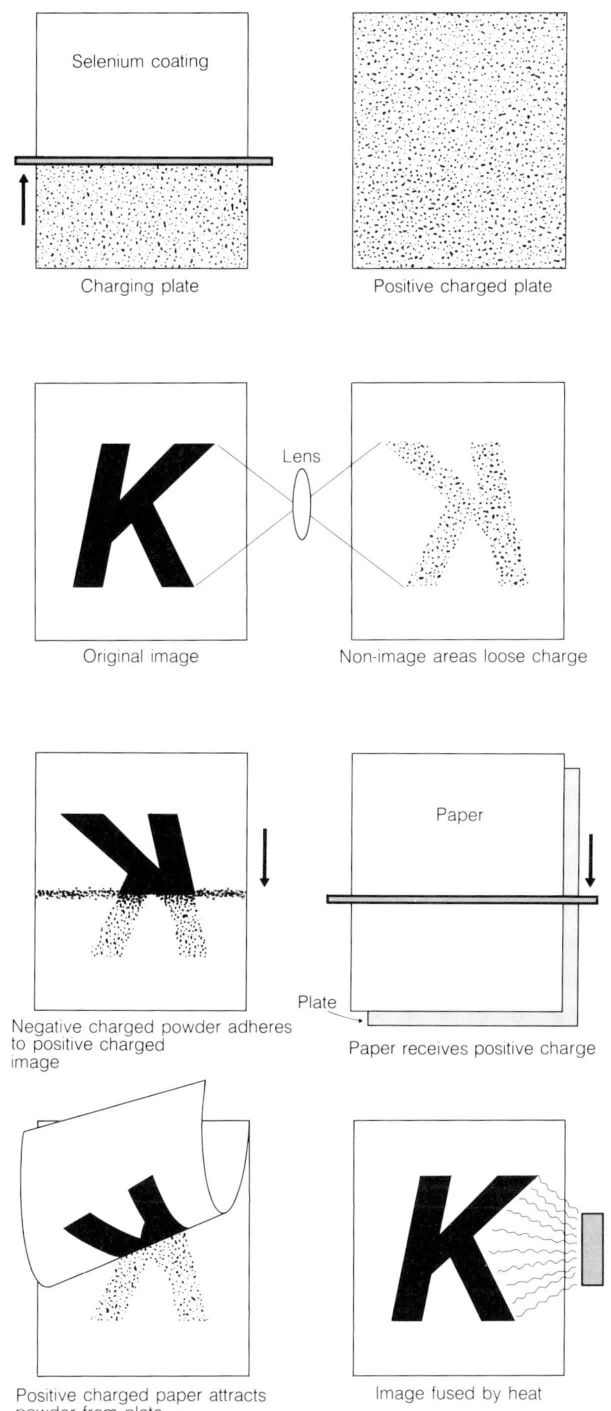

Fig. 10.11 Indirect electrostatic process

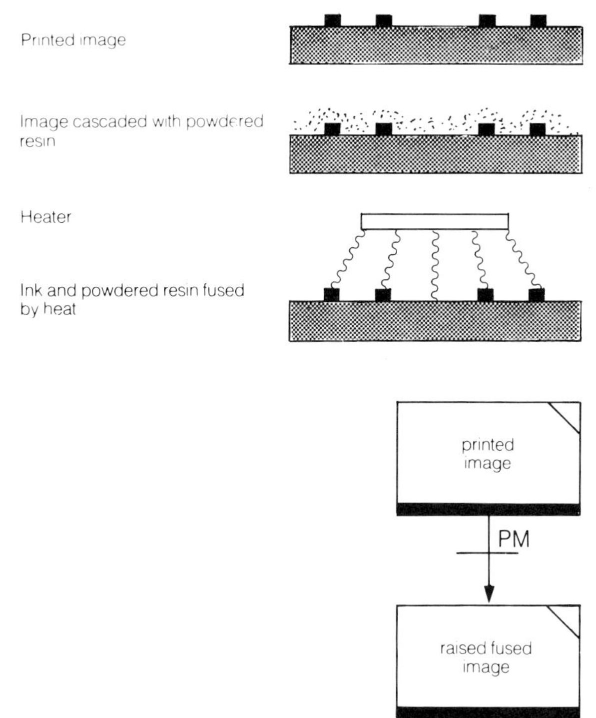

Fig. 10.12 Thermograph process for producing raised images

10.7 Thermography

In this process, the printed sheets are usually delivered directly from the printing press to a moving belt which carries them through a thermograph machine in which they are cascaded with a resin-like powder that adheres to the ink (Fig. 10.12). The sheets are then passed through a heater which fuses the ink and powder causing the image to be raised slightly above the surface of the printed material.

Some of the products produced by this method include, business cards, invitations, personal stationery, greeting cards, etc. Tactual maps for the visually impaired can also be produced by this process by using specially prepared inks that will raise to a much greater extent when subjected to heat.

References and Bibliography

Albertsson, Arne & Sahlstrom, Rolf. 1969, *Matningsteknik*, Harnoforlaget, Harnosand.

Bem, S. 1960, *Rysunek map*, Panstwowe Przedsiebiorstwo Wydawnictw Kartograficznych (PPWK), Warszawa, Polska, 258 pp. 1966, 391 pp.

Bem, S. 1968, *Wzyory pism*, PPWK, Warszawa, Polska, 22 pp.

Boczar, S. 1982, *Kartografia. Opracowanie map i reprodukcja kartograficzna*, Wydawnictwa Akademii Gorniczo – Hutniczej (Wyd. AGH), Krakow, Polska.

Birkstedt, Seppo. 1970, *Teemakarttojen kuvausperusteista ja laatimismenettelysta*, diplomityo, TKK Maanmittausosasto, Helsinki, Finland, 112 pp.

Blomberg, Leo. 1976, Automaattisen piirustuskojeen avulla tuotettavista teemakartoista, *Maankaytto* n:o 1.

Boratynski, K. & Kwintkiewicz, M. 1984, *Maszyny i urzadzenia reprodukcyjna*, Wydawnictwa Szkolne i Pedagogiczne, Warszawa, Polaska, 212 pp.

Brokman, L. & Kolanowski, S. 1972, *Poradnik technika pracowni kartograficzno-reprodukcyjnej*, PPWK, Warszawa, Polska, 328 pp.

Carlsson, Gosta E. 1967, *Grafisk Teknik*, Grafiska Forskningslaboratoriet (GFL), Stockholm.

Cichowska-Cieslak, H. & Makowski, A. 1962, *Kurs cwiczen z reprodukcji kartograficznej*, Wydawnictwa Politechniki Warszawakiej (WPW), Warszawa, Polska, 329 pp., 1972, 251 pp.

Ciesielski, J. 1969, *Metody reprodukcji malonakladowej*, Wydawnictwo Instytutu Geodezji i Kartografii, (IGiK), Warszawa, Polska.

Cuenin, Rene, 1973, *Cartographie generale* Collection scientifique de l'Institut Geographique National, Editions Eyrolles, 203 pp.

Donell, Jr, Porter, W. & Dekker, AG, Marcel. 1979 *Introduction to Map Projections*, CH–4010 Basel.

Elsner, B. & Bester, J. 1986, *Fotografia reprodukcyjna*, WPW, Warszawa, Polska, 265 pp.

Elvhage, Christian, 1983, *Terrangatergivning, En serie svenska experimentkartor*, UNGI Rapport nr. 59, Uppsala Universitet, Naturgeografiska Inst., Box 554, 751 22 Uppsala.

Fargboken, Svenska Slojdforeningen, Raben och Sjogren, Stockholm, 1965.

Forskningsarkiv och kartlagning, utredningsrapport Exp A 249, Dnr 14–111–77, Statens Lantmateriverk, 1977.

Grafisk teknik idag, Grafiska Forskningslaboratoriet (GFL), Stockholm, 1973.

Graphics Arts Technological Foundation Inc. 1980, Raymond Blair & Charles Shapiro, Editors; *The Lithographers Manual*, 6th edition, Pittsburgh, Pennsylvania, USA.

Gruszczynski, Cz. 1984, *Farby graficzne*, Wydawnictwa Szkolne i Pedagogiczne, Warszawa, Polska, 160 pp.

Grygorenko, W. 1970, *Radakcja i opracowanie map ogolnogeograficznych*, PPWK, Warszawa, Polska, p. 347–357.

Grygorenko, W. 1973, *Reprodukcja kartograficzna*, Wydawnictwa Wojskowa Akademii Technicznej, Warszawa, Polska, 177 pp.

Guethner, T. 1978, *Fotografia techniczna*, WPW, Warszawa, Polska, 194 pp.

Guethner, T. 1980, *Podstawy fotografii*, PPWK, Warszawa, Polska, 180 pp.

Glowny Urzad Geodezji i Kartografii. 1983, *Wytyczne techniczne K-1.4: Mapa zasadnicza*, Wydawnictwa Glownego Urzedu Geodezji i Kartografii, Warszawa, Polska, 18 pp.

Glowny Urzad Geodezji i Kartografii. 1981, *Wytyczne techniczne K-1.7: Mapa zasadnicza w wersji rozwarstwicnej*, Wydawnictwa Glownego Urzedu Geodezji i Kartografii, Warszawa, Polska, 28 pp.

Glowny Urzad Geodezji i Kartografii. 1981, *Wytyczne techniczne K-1.9: Sporzadzanie pierworysu mapy. Materialy kartograficzne. Sprzet kreslarski, tusze, technika rysowania*, Wydawnictwa Glownego Urzedu Geodezji i Kartografii, Warszawa, Polska, 88 pp.

Glowny Urzad Geodezji i Kartografii. 1981, *Wytyczne techniczne K-1/K-3.10. Reprodukcja kartograficzna malonakladowa. Metody i technologie*, Wydawnictwa Glownego Urzedu Geodezji i Kartografii, Warszawa, Polska, 84 pp.

Glowny Urzad Geodezji i Kartografii. 1983, *Wytyczne techniczne K-2.3: Sporzadzanie map fotograficznych*, Wydawnictwa Glownego Urzedu Geodezi i Kartografii, Warszawa, Polska, 90 pp.

Haftman, H. 1983, *Podstawy techniki pomiarowej dla poligrefow*, Wydawnictwa Naukowo – Techniczne (WNT), Warszawa, Polska, 102 pp.

Hajek, M. 1978, *Kartograficka tvorba a reprodukcia*, (Map Production and Reproduction) (in Slovak) Slovenska vysoka skola technicka, Bratislava, Czechoslovakia, 330 pp.

Hallberg, Ake. *Grafisk Design*, Bokforlaget Spektra, Halmstad.

Hallberg, Ake. 1977, *Klart for tryck*, Bokforlaget Spektra, Halmstad.

Hallberg, Ake, 1967, *Trycksaksbestallning*, Wahlstrom & Widstrand, Stockholm.

Hallberg, Ake. 1969, *Typografi*, Bokforlaget Spektra, Halmstad.

Heden, Stig J. 1982, *Grafiska ord i dataaldern*, Liber/ Allmanna Forlaget, Stockholm.

Hesse, H.J. 1975, *Cartographic Generalization*, c/o Orell Fussli Graphische Betriebe, Postfach, CH-8036 Zurich.

Imhof, Eduard. 1968, *Gelande und Karte*, Eugen Rentsch Verlag, Erlenbach-Zurich.

Imhof, Eduard. 1965, *Kartographische Gelandedarstellung*, Walter de Gruyter & Co., Berlin 30.

International Cartographic Association. 1984, *Basic Cartography for Students and Technicians*, ICA with U.N.E.S.C.O. assistance.

International Cartographic Association. 1980, *Colourproofing in Cartography*, Commission 2, ICA, (in binder).

International Yearbook of Cartography, 1961–1978 osv, Kirschbaum Verlag, Bonn-Badbodesberg.

Ivarsson, Ivan. 1973, *Fotosattningsmaskiner och system for textproduktion*, Grafiska Forskningslaboratoriet (GFL), Stockholm.

Kartografiska Sallskapet. 1978, *Sveriges Kartlaggning*, Liber Kartor, Stockholm, Forsta urgava, 1922, tillagg vart 10:e ar, senast.

Kartor forr och nu, (oversattning), Wahlstrom & Widstrand, Stockholm, 1971.

Keates, J.S. 1973, *Cartographic Design and Production*, Longman, London, England, 204 pp.

Keates, J.S. 1982, *Understanding Maps*, Longman, London, England, 139 pp.

Keisteri, Tapio. 1981, *Kartografien yleistaminen, automaation mahdollisuudet*, Luentomoniste, TKK Maanmittausosasto, Helsinki, Finland, 12 pp.

Kers, A.J. 1980, *Review of Registration Systems*, ITC Bookshop, Enschede, Netherlands, 26 pp.

Klinghammer, I. & Papp-Vary, A. 1983, *Foldunk tukre a terkep* (The map as a mirror of the Earth), Gondolat, Budapest, Hungary, 385 pp.

Kovarik, J. & Veverka, B. 1980, *Kartograficka tvorba* (Map Production), Publishing House CVUT, Praha, Czechoslovakia, 180 pp.

Kraus, V. 1980, *Kartoreprodukce* (Map Reproduction), Publishing House CVUT, Praha, Czechoslovakia, 212 pp.

Kwasnik, L. & Nowicka, H. 1975, *Cwiczenia laboratoryjne z materialoznastwa poligraficznego*, WPW, Warszawa, Polska, 249 pp.

Lawrence, G.R.P. 1979, *Cartographic Methods*, Methuen & Co. Ltd., London, England, 162 pp.

Lehto, E. 1981, *Digitaalinen tulkittu Landsat-aineisto alueellisen tietojarjestelman osana*, diplomityo, TKK Maamittausosasto, Helsinki, Finland, 70pp.

Loxton, John. 1980, *Practical Map Production*, John Wiley & Sons, Chichester, England.

Maanmittaushallitus, Suomen Maantieteellinen Seura. 1984 Suomen Kartasto, vihko 112, *Suomen Kartoitus* (Atlas of Finland, Folio 112, *Mapping of Finland)*, Maanmittaushallitus, Helsinki, Finland, 40 pp.

Maanmittaushallitus. 1985, *Kopiotekniikan ohjeisto*, Maanmittaushallitus, Helsinki, Finland, 100 pp.

Makkonen Kirsi. 1986, *Perustietoa tietokoneavusteisesta kartografiasta*, kasikirjoitus, TKK Maanmittausosasto, Helsinki, Finland, 91 pp.

Maling, D.H. 1973, *Coordinate Systems and Map Projections*, George Philip & Son, London.

Makowski, A. 1976, *Podstawa technologii barwy w kartografii*, WPW, Warszawa, Polska, 108 pp.

Meyen, E. 1973, *Multilingual Dictionary of Technical Terms in Cartography*, Steiner Verlag GmbH, Wiesbaden.

Miksovsky, M. 1987, *Kartografie* (Cartography), Geodeticky a kartograficky podnik, Praha, Czechoslovakia, 209 pp.

Ministerstwo Rolnictwa. Department Urzadzen Rolnych. 1973, *Wytyczna wykonywania prac kartograficzno-reprodukcyjnych* (cz. I, II, III), Wydawnictwa Ministerstwa Rolnictwa, Warszawa, Polska.

Modern Grafisk Teknik, del I, Grafiska Forskningslaboratoriet (GFL), Stockholm, 1980.

Osowski, F. & Brokman, L. 1970, *Elementy kartografii*, PPWK, Warszawa, Polska, 179 pp.

Palm, Christer & Lofgren, Kurt. 1985, *Compendium in Cartography*, Geodesi Royal Institute of Technology, Stockholm, Sweden, 300 pp.

Piatkowski, F. 1953, *Kartografia i reprodukcja kartograficzna*, PPWK, Warszawa, Polska, 406 pp.

Piatkowski, F. 1957, *Fototechnika w reprodukcji kartograficznej*, PPWK, Warszawa, Polska, 169 pp.

Piatkowski, F. 1969, *Kartografia, redakcja map i reprodukcja kartograficzna*, Panstwowe Wydawnictwa Naukowe (PWN), Warszawa, Polska, 289 pp.

Piatkowski, F. & Olejniczak, L. 1978, *Specjalne technologie reprodukcji*, WPW, Warszawa, Polska, 138 pp.

Raisz, Erwin Josephus, 1962. *Principles of Cartography*, McGraw-Hill, New York, USA, 315 pp.

Robinson, Arthur H., Sale, Randall & Morrison, Joel. 1978, *Elements of Cartography*, 4th Edition, John Wiley & Sons, New York, USA, 448 pp.

Sasvari, E. 1975, *Kartografia ismeretek* (Basics of Cartography), Cartographia, Budapest, Hungary, 157 pp.

School of Military Survey. 1971, *Military Engineering*, Vol. XIII-Part XII, Cartography, Ministry of Defence, London, England, 204 pp.

School of Military Survey. 1967, *Military Engineering*, Vol. XIII-Part XIII, Map Reproduction, Ministry of Defence, London, England, 241 pp.

Svenska Kartor, Svenska Turistforeningen, publikation 2345, 1979.

Szaflarski, J. 1965, *Zarys kartografii*, PPWK, Warszawa, Polska, 699 pp. Ternryd, Carl-Olof & Lundin, Eliz. 1966, *Matningsteknik och Fotogrammetri*, Akademiforlaget Gumperts, Goteborg.

Toivonen, Esa. 1983, *Kunnallistekniset suunnitelmat ja mittaukset 1*, Ammattikasvatushallitus, Valtion Painatuskeskus, Helsinki, Finland, 83 pp.

Tuhkanen, Timo. 1984, *Topografinen kartografia*, Luentomoniste, TKK Maanmittausosasto, Helsinki, Finland, 320 pp.

Vahala, Matti. 1974, Eraiden sovellettujen karttojen valmistuksen automatisointi, *Suomen kunnallislehti* n:o 7.

Wilhelmy, Herbert. 1966, *Kartographie in Stichworten*, Verlag Ferdinand Hirt, Kiel.

Weitt, Werner. 1967, *Thematische Kartografie*, Gebruder Janecke Verlag, Hannover.

Ziernow, W.A. 1972, *Procesy fotografgiczne w technice reprodukcyjnej*, WNT, Warszawa, Polska, 377 pp.

PERIODICALS

Bulletin of the Geographical Survey Institute, Geographical Survey Institute, Japan.

Cartographica, B.V. Gutsell, Editor, University of Toronto Press, Toronto, Canada.

'*Chizu*', Japan Cartographers Association, Tokyo, Japan.

ITC Journal, Ann Stewart, Editor, International Institute for Aerial Survey & Earth Sciences (ITC), Enschede, Netherlands.

The Cartographic Journal, G.R.P. Lawrence, Editor, The British Cartographic Society, London, England.

The American Cartographer, A.Jon Kimmerling, Editor, American Congress on Surveying & Mapping, Falls Church, Virginia, USA.

Index